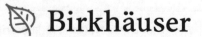

Progress in Nonlinear Differential Equations and Their Applications

PNLDE Subseries in Control

Volume 99

More information about this subseries at https://link.springer.com/bookseries/15137

Alexandre Bayen · Maria Laura Delle Monache
Mauro Garavello · Paola Goatin · Benedetto Piccoli

Control Problems for Conservation Laws with Traffic Applications

Modeling, Analysis, and Numerical Methods

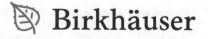 Birkhäuser

Alexandre Bayen (ID)
Institute of Transportation Studies
University of California
Berkeley, CA, USA

Maria Laura Delle Monache
Department of Civil and
Environmental Engineering,
Part of the work was done while the author
was at Inria Grenoble-Rhône Alpes (France)
University of California, Berkeley
Berkeley, CA, USA

Mauro Garavello
Department of Mathematics & Applications
University of Milano-Bicocca
Milan, Italy

Paola Goatin
Inria, Research Center of Université Côte
d'Azur
Sophia Antipolis, France

Benedetto Piccoli
Department of Mathematical Sciences
Rutgers University
Camden, NJ, USA

http://dx.doi.org/10.13039/100006978, University of California, Berkeley

This title is freely available in an open access edition with generous support from the Library of the University of California, Berkeley.

ISSN 1421-1750   ISSN 2374-0280   (electronic)
Progress in Nonlinear Differential Equations and Their Applications
ISSN 2731-7366   ISSN 2731-7374   (electronic)
PNLDE Subseries in Control
ISBN 978-3-030-93017-2   ISBN 978-3-030-93015-8   (eBook)
https://doi.org/10.1007/978-3-030-93015-8

Mathematics Subject Classification: 35L65, 35Q93, 49J20, 93B05, 93B52, 60K30, 90B20

This book is published under the imprint Birkhäuser, www.birkhauser-science.com by the registered company Springer Nature Switzerland AG
The registered company address is: Gewerbestrasse 11, 6330 Cham, Switzerland

To my wife Zoë for her ongoing love and support.

To my grandmothers Maria and Rita.

To my grandparents.

To Marcella and Carlo, my parents.

To my siblings Gregorio, Antonio, and Maria and to the memory of Giovanni.

Contents

List of Figures

List of Tables

Chapter 1
Introduction

This book focuses on control problems for conservation laws, i.e., equations of the type:

$$\partial_t u + \partial_x f(u) = 0 \qquad u_t + (f(u))_x = 0, \qquad (1.1)$$

where $u : \mathbb{R}^+ \times \mathbb{R} \to \mathbb{R}^n$ is the vector of conserved quantities and $f : \mathbb{R}^n \to \mathbb{R}^n$ is the flux. Most results will be given for the scalar case ($n = 1$), but we will present few results valid in the general case.

We consider four types of control problems and use ω to denote the control variables. Namely:

- **Boundary control**: We restrict (1.1) to $x \in [a, b]$ and control the boundary values: $u(t, a) = \omega_a(t)$, $u(t, b) = \omega_b(t)$.
- **Decentralized control**: We consider (1.1) on a network and control distribution parameters at nodes.
- **Distributed control**: We assume to control some parameters defining the flux, thus $f = f(u, \omega)$.
- **Lagrangian control**: We assume the flux depends on the position of N controlled *particles, $f = f(u, y)$, $y = (y_1, \ldots, y_N)$* and $\dot{y}_i = \omega_i$.

The interest in conservation laws and their control is motivated by a large and diverse collection of applications. While classical fluid dynamic problems (also covering different areas) motivated research since more than hundred years, a new set of applications motivated recent interests. Many of the latter include problems formulated on networks, which are represented by topological graphs [142]. Among many, let us mention the following: vehicular traffic, water canals, supply chains, air traffic control, data networks, and service networks (gas, water, electricity, etc.).

We give particular attention to vehicular traffic modeling. Classical control problems in this domain correspond to traffic regulation at fixed locations (such as traffic lights, traffic signals, pay tolls, etc.), while the advent of autonomy

© The Author(s) 2022

A. Bayen et al., *Control Problems for Conservation Laws with Traffic Applications*,
PNLDE Subseries in Control 99, https://doi.org/10.1007/978-3-030-93015-8_1

and communication has opened the possibility of more distributed and ubiquitous controls. More precisely, boundary controls correspond to entrance points and tolls and decentralized controls to traffic signals at junctions. On the other side, variable speed limit gives rise to distributed control and use of autonomy and communication to Lagrangian control problems. Summarizing, vehicular traffic presents potential applications for controls in all the four categories mentioned above. It is well known that conservation laws are strictly connected to Hamilton-Jacobi equations, thus we include a chapter on control of the latter. Also in this case, vehicular traffic applications were among the strongest motivation for researchers.

Many books have been devoted to conservation laws and to their control. Let us mentions the following by categories:

General theory of conservation laws: [51, 99, 213, 236, 251, 252];

Control problems for hyperbolic equations: [32, 103];

Hyperbolic equations on networks: [103, 233];

Hamilton-Jacobi equations: [27, 126, 225];

Modeling of vehicular traffic: [139, 142, 230].

The book can be used for a one semester course at graduate or advanced undergraduate level. The undergraduate students should have been previously exposed to Partial Differential Equations. However, since most of the materials are based on conservation laws, we included Appendix A dealing with the general theory of initial-boundary values problems for balance laws (i.e., including possible of source terms). Readers which are not expert in conservation laws may also want to use as references the textbooks: [51, 99, 251, 252]; On the other side, the theory of conservation laws on networks, i.e., topological graphs, was more recently developed, thus we included Appendix B illustrating the main concepts. The latter are then thoroughly investigated in Chap. 3 to deal with control problems. There are at least three possible course design:

1. Traffic modeling using control of conservation laws. This course would be for investigators more interested in the applications to traffic. Course material: Chap. 1, Appendix B, Chap. 2 (Sects. 2.1, 2.3.2, 2.3.4, 2.4), Chap. 3 (Sects. 3.1, 3.3, 3.4, 3.6), Chap. 4 (Sects. 4.2–4.4), Chap. 5.
2. Mathematical control theory for balance laws. This is for researchers more interested in the mathematical aspects of control. Course material: Chap. 1, Appendix A, Chap. 2 (Sects. 2.1–2.3), Chap. 3 (Sects. 3.1–3.2), Chap. 4 (Sect. 4.1), Chap. 5 (Sects. 5.1–5.3), Chap. 6.
3. Conservation laws on networks and control problems. This is for researchers interested in the general theory of conservation laws on networks and their application. Course material: Chap. 1, Appendix A, Appendix B, Chaps. 2, 3 (Sects. 3.1–3.5), Chap. 4 (Sects. 4.1–4.2), Chap. 6.

The book is organized as follows.

Chapter 2 deals with boundary control problems. We first briefly summarize results for the case of solutions with no shocks (Sect. 2.2). Then illustrate general results for attainable sets (Sect. 2.3), Lyapunov techniques for the scalar case with two boundaries (Sect. 2.4), finally mixed PDE-ODE systems (Sect. 2.5).

Chapter 3 is focused on decentralized controls. These controls act, for instance, at nodes of a network by regulating fluxes. A perfect example is that of traffic lights, ramp metering, and pay tolls. The control problem is formulated in terms of the Riemann solver at nodes (Sect. 3.2), then focusing on signalized junctions (Sect. 3.3), freeway control (Sect. 3.4), and inflow control (Sect. 3.5). We also consider the optimization of travel times and emergency management on networks (Sect. 3.6).

Chapter 4 considers distributed control for conservation laws. The stability of Riemann solvers is a key ingredient to deal with control problems (Sect. 4.1). Our main application is variable speed limit, with results on general control problems (Sect. 4.2), discrete optimization (Sect. 4.3), and systems of equations (Sect. 4.4).

Chapter 5 illustrates Lagrangian control problems. First we introduce coupled ODE-PDE models for moving bottlenecks (Sect. 5.2) and then we develop numerical methods (Sect. 5.3). Applications to traffic management are illustrated both numerically and experimentally (Sect. 5.4).

Chapter 6 explores relations between conservation laws, Hamilton-Jacobi equations, and their control. First strong solutions are considered (Sect. 6.2), then generalized ones (Sect. 6.3). General optimization problems are then discussed (Sect. 6.4).

Appendix A provides a brief introduction to initial and boundary value problems for conservation and balance laws, while Appendix B focuses on conservation laws on networks to model traffic.

The authors wish to thank J.-M. Coron for encouragement during the whole writing process. They are also thankful to Amaury Hayat for contributing to the second chapter and to Alexander Keimer and Christian Claudel for contributing to the sixth chapter.

This work was supported by the following funding: National Science Foundation Cyber-Physical Systems Synergy Grant No. CNS-1837481 and the endowment of the Joseph and Loretta Lopez Chair (B.P.); the Liao-Cho Chair (A.B.). The authors are also thankful to their families for the continuing support and patience during the working of the book.

Chapter 2
Boundary Control of Conservation Laws Exhibiting Shocks

2.1 Introduction

This chapter focuses on control of systems of conservation laws with boundary data. Problems with one or two boundaries are considered and, in particular, we focus on cases where shocks may be developed by the solution. However, for completeness we briefly discuss in Sect. 2.2 other existing results where singularities are prevented via suitable feedback controls such as in [32].

More precisely, let us consider the system of conservation laws

$$\partial_t u + \partial_x f(u) = 0, \tag{2.1}$$

where the unknown u is defined in the set $\mathcal{D} = \{(t, x) \mid t \geq 0 \text{ and } a \leq x \leq b\}$, $a \in \mathbb{R}$ and $b \in \mathbb{R} \cup \{+\infty\}$, and has values on $\Omega \subseteq \mathbb{R}^n$, with $n \geq 1$. The flux function $f : \Omega \to \mathbb{R}^n$ is assumed to be smooth (infinitely differentiable) and each characteristic field to be genuinely nonlinear or linearly degenerate (see Definition 45 in Appendix A). The initial-boundary value problem (IBVP) for (2.1) with initial condition $u_0 : (a, b) \mapsto \Omega$, and boundary controls $\omega_a, \omega_b : \mathbb{R}^+ \mapsto \Omega$, reads

$$\partial_t u + \partial_x f(u) = 0, \tag{2.2}$$

$$u(0, x) = u_0(x), \tag{2.3}$$

$$u(t, a) = \omega_a(t), \quad u(t, b) = \omega_b(t). \tag{2.4}$$

For basic theory and well-posedness results for system (2.2)–(2.4), we refer the reader to Appendix A.8.

© The Author(s) 2022
A. Bayen et al., *Control Problems for Conservation Laws with Traffic Applications*,
PNLDE Subseries in Control 99, https://doi.org/10.1007/978-3-030-93015-8_2

2.2 Boundary Controls for Smooth Solutions *Co-authored by Amaury Hayat*

Many previous studies exist on boundary control of conservation laws for regular solutions, not presenting shocks. The problem of finding a boundary control to stabilize a steady state of nonlinear conservations laws goes back to [255] and [158]. In the latter, J. Greenberg and T. Li are studying carefully the $\mathbf{C}^1$ solutions along the characteristics of two coupled conservation laws. These results were later extended for a general system of conservation laws in [220] and by T.-H. Qin in [245]. They consider strictly hyperbolic systems of the form (2.1) where all the eigenvalues of $Df(0)$ are non-vanishing and where the boundary controls are output feedbacks, meaning that they depend only on the output information of the system. Under the assumption that the solution is of class $\mathbf{C}^1$ it can be assumed without loss of generality that all the eigenvalues of $Df(0)$ are positive and the output feedbacks take the generic form

$$u(t, a) = G(u(t, b)), \tag{2.5}$$

where G is the feedback function, also assumed to be $\mathbf{C}^1$. The result they show is the following

Theorem 1 *The system* (2.1) *with boundary condition* (2.5) *is (locally) exponentially stable for the* $\mathbf{C}^1$ *norm provided*

$$\rho_\infty(G'(0)) = 1. \tag{2.6}$$

Here the quantity ρ_∞ is defined by

$$\rho_p(M) = \min\left\{ \|\Delta M \Delta^{-1}\|_p : \Delta \in \mathcal{D}_n^+ \right\}, \quad \text{for } p \in \mathbb{N}^* \cup \{+\infty\}, \tag{2.7}$$

where $\mathcal{D}_n^+$ denotes the space of diagonal matrix with positive coefficients and $\|M\|_p$ refers to the matrix norm $\sup_{\xi \in \mathbb{R}^n} \|M\xi\|_p / \|\xi\|_p$. Condition (2.6) is only sufficient, and there is a gap between this condition and the stability condition of the associated linear system. For the latter, a necessary and sufficient condition for an exponential stability robust with changes in the propagation speed was shown to be (see [164, Theorem 6.1])

$$\rho_0(G'(0)) < 1, \quad \text{where} \tag{2.8}$$

$$\rho_0(G'(0)) = \max\{\rho(\text{diag}(e^{i\theta_1}, \ldots, e^{i\theta_n})G'(0)) : \theta_i \in \mathbb{R}\}. \tag{2.9}$$

Here $\rho(M)$ denotes the spectral radius of M, for $M \in M_n(\mathbb{R})$, and exponential stability robust with changes in the propagation speed mean that there exists $\varepsilon > 0$ such that, for any $D\hat{f}(0)$ satisfying $|D\hat{f}(0) - Df(0)| < \varepsilon$, the system

$$\partial_t u + D\hat{f}(0)\partial_x u = 0, \tag{2.10}$$

with boundary conditions (2.5) is exponentially stable. Besides, $\rho_0 \leq \rho_\infty$ and the two quantities do not coincide in general. Other studies using a Lyapunov approach recovered later the same stabilization result in the $\mathbf{C}^1$ norm but also extended it to the H^q norms for $q \geq 2$ and to various settings [32, 87, 88, 90, 110]. In particular

Theorem 2 *The system* (2.1) *with boundary condition* (2.5) *is exponentially stable in the* H^2 *norm if*

$$\rho_2(G'(0)) < 1. \tag{2.11}$$

This is interesting as $\rho_2(G'(0)) \neq \rho_\infty(G'(0))$ in general, but also $\rho_2(G'(0)) = \rho_0(G'(0))$ as long as $n \leq 5$ (see [88]). To show these results, they use Lyapunov function candidates of the form of weighted norms of the solution and its time derivatives[1]

$$W_{p,q}(t) = \left(\sum_{k=0}^{q} \int_a^b \sum_{i=1}^n p_i^p (\partial_t^k u_i(t,x))^{2p} e^{-2\mu px} \, dx \right)^{1/2p} dx. \tag{2.12}$$

For the H^2 norm, one can directly use $W_{1,2}$, as this Lyapunov function candidate is directly equivalent to the H^2 norm. For the $\mathbf{C}^1$ norm, one has first to find some estimates on $W_{p,1}$ that have to be uniform on p provided that p is large enough. Then by letting p go to $+\infty$ one recovers a quantity equivalent to the $\mathbf{C}^1$ norm. In both cases, the question of the stabilization reduces to finding a sufficient condition on the coefficients p_i and μ such that the Lyapunov function candidate decreases exponentially along the solutions of the system and in a distributional sense.

Theorem 2 was later extended by J.-M. Coron and H.-M. Nguyen to the $\mathbf{W}^{2,p}$ norm with ρ_p instead of ρ_2 in the sufficient condition (2.11), by using a time delay approach [93]. But, more importantly, they showed that specifying the norm is not superfluous: a nonlinear system could be exponentially stable in the H^2 norm and not exponentially stable in the $\mathbf{C}^1$ norm. More precisely, they showed the following

Theorem 3 *Let $n \geq 2$ and $\tau > 0$, there exists $f \in \mathbf{C}^\infty(\mathbb{R}^n; \mathbb{R}^n)$ such that $Df(u)$ is diagonal and $Df(0)$ has distinct positive eigenvalues, and a linear feedback $G :$ $\mathbb{R}^n \to \mathbb{R}^n$ such that*

---

[1] Actually, as we are considering solutions of the conservation laws (2.1), the Lyapunov function candidates can be expressed only as a function of the solution and its space derivatives. The expression is then more complicated but it illustrates that the Lyapunov function can be seen as a functional on functions of one variable, just like the $\mathbf{C}^1$ or H^2 norm.

$$\rho_\infty(G'(0)) < 1 + \tau,$$
$$\rho_2(G'(0)) = \rho_0(G'(0)) < 1, \tag{2.13}$$

and the system (2.1), (2.5) *is **not** exponentially stable in the* $\mathbf{C}^1$ *norm.*

This result implies that for conservations laws, in contrast with finite-dimensional systems, the stability of the linearized system does not necessarily imply the stability of the nonlinear system. This explains the gap between the linear condition $\rho_0(G'(0)) < 1$ and (2.6). It can further be showed that one always have $\rho_2(G'(0)) \le \rho_\infty(G'(0))$. This, and the simpler expression of the Lyapunov function for the H^2 norm compared to the $\mathbf{C}^1$ norm, explains that several particular systems of conservation laws were studied in the H^2 (or H^p) norm framework (see, e.g., [33, 120, 162, 171]). More practical controllers, like proportional integral controllers, where also considered, for instance, in [269], or in [92]. In the latter, the authors apply such controller to a scalar conservation law. They obtain a necessary and sufficient stability condition by extracting from the solution the part that limits the stability, using a suitable projector. Then, they conclude by studying carefully this projection, while using a Lyapunov function for the remaining part of the solution.

For balance laws, the previous results can be generalized but the situation is intrinsically more complex. Indeed, the steady states to be stabilized are not necessarily uniform, and the source term can strongly couple the equations: even the linearized system cannot anymore be written as independent equations where the coupling only comes from the boundary. To deal with this issue, in [32, Chapter 6] and in [170], the authors used Lyapunov functions similar to (2.12) but where the coefficients p_i are now replaced by space-dependent weights $f_i(x)$. A boundary condition similar to (2.6) or (2.11) appears but it turns out that another condition, independent of the boundary control and intrinsic to the system, also appears.

It is worth mentioning that more general controls were proposed for balance laws in order to solve the issues related to the source term. One can cite, for instance, the backstepping approach. Taking its name from a method used on finite-dimensional systems, the backstepping approach was adapted to PDEs in [89] and then modified in [25, 46, 197]. It consists in finding an invertible application to map the balance laws to a simpler system, usually conservation laws without source term. Then, it is possible to deduce a control for the balance laws by finding a control on the simpler system and using this invertible application. The consequence of such strategy is that the control is often a full-state feedback. This method was first used for hyperbolic system in [118, 180, 196] by applying a Volterra transform of the second kind, namely a transform of the following form

$$(u(t, x)) \longmapsto \left(u(t, x) - \int_a^x K(x, y) u(t, y) \, \mathrm{d}y \right). \tag{2.14}$$

In all the above cases, however, the boundary control ensures that the solution will remain of class $\mathbf{C}^1$ or H^2, provided the initial data is itself $\mathbf{C}^1$ or H^2. So, shocks can never form. When one wants to deal with solution including shocks or discontinuous initial data, none of these results can be applied and much less is known. One can cite [35] where the authors aim at stabilizing a shock steady state for a scalar equation, from an initial data with a single shock and regular otherwise. To do so, the authors consider the solution with shock as two regular functions, one before the shock and one after the shock, coupled with each other by the boundary conditions and the dynamics of the shock given by the Rankine–Hugoniot conditions. The problem becomes equivalent to two conservations laws coupled with an ODE and the goal is to stabilize at the same time both the conservations laws and the ODE, and this is done defining a kind of hybrid Lyapunov function. The sufficient stability condition they obtain is an analogous of (2.11). Also, in [241] the author deals with discontinuous solutions in the **BV** class. This approach is closer to the one presented in this book, and the results is quite powerful, but it requires not only a boundary control but also an internal control. Finally, in [91] the authors aim at stabilizing a null steady state for two coupled conservation laws starting from potentially discontinuous solutions. More precisely they show the following

Theorem 4 *Let the system (2.1) be strictly hyperbolic and genuinely nonlinear, assume that the velocities are positive, and that G is linear. Assume in addition that*

$$
\inf_{\alpha \in (0,\infty)} \Big\{ \max \Big\{ |l_1(0).G'(0)r_1(0)| + \alpha |l_2(0).G'(0)r_1(0)|,
$$
$$
|l_2(0).G'(0)r_2(0)| + \alpha^{-1}|l_1(0).G'(0)r_2(0)| \Big\} \Big\} < 1,
\tag{2.15}
$$

where l_1, l_2 and r_1, r_2 are respectively the left and right eigenvectors of $Df(u)$ corresponding to the eigenvalues $\lambda_1(u) > 0$ and $\lambda_2(u) > 0$.

Then there exists $\varepsilon > 0$, $C > 0$ and $\gamma > 0$ such that for any $u_0 \in \mathbf{BV}(0, L)$ with $\|u_0\|_{\mathbf{BV}} \leq \varepsilon$, there exists an entropy solution in $\mathbf{L}^\infty(0, \infty; \mathbf{BV}(0, L))$ to the system (2.1) with initial condition u_0 and satisfying (2.5) for almost all time, and such that

$$
\|u(t, \cdot)\|_{\mathbf{BV}} \leq Ce^{-\gamma t}\|u_0\|_{\mathbf{BV}}, \quad \text{for all } t \geq 0.
\tag{2.16}
$$

However, the result has some limitations: there exists a solution which converges exponentially in the **BV** norm, but there are no guarantees on its uniqueness. Besides, the velocities have to be positive meaning that $Df(u)$ has to be definite positive, which can be assumed without loss of generality when the solutions are $\mathbf{C}^1$, but restrict the cases when the solutions present shocks.

2.3 The Attainable Set

Aim of this section is to characterize the attainable set for the initial-boundary valued problem (2.2), (2.3), and (2.4), i.e., the set of the profiles which can be attained at a fixed time T for a fixed initial datum $u_0 \in \mathbf{L}^1(a, b) \cap \mathbf{L}^\infty(a, b)$. More precisely, given two sets $\mathcal{U}_a, \mathcal{U}_b \subseteq \mathbf{L}^\infty(0, T)$, let us define the attainable set

$$\mathcal{A}(T, \mathcal{U}_a, \mathcal{U}_b, u_0) = \{u(T, \cdot) : u \text{ sol. to } (2.2) - -(2.4) \text{ with } \omega_a \in \mathcal{U}_a, \ \omega_b \in \mathcal{U}_b\}. \tag{2.17}$$

We remark that conservation laws, in general, generate discontinuities in finite time in the solution even if the initial and boundary conditions are smooth. The space of bounded variation functions represents the correct setting for solutions. Hence the set $\mathcal{A}(T, \mathcal{U}_a, \mathcal{U}_b, u_0)$ is a subset of $\mathbf{BV}(a, b)$.

2.3.1 The Scalar Case with a Single Control

In this section, we consider the conservation law (2.1) on the domain $\mathcal{D} = (0, T) \times (0, +\infty)$, with $T > 0$ fixed, $n = 1$, $\Omega = \mathbb{R}$, $a = 0$, and $b = +\infty$. The flux function f is assumed to be smooth (infinitely differentiable) and strictly convex; the strictly concave case is entirely similar. The initial-boundary value problem in this situation for (2.1) reads

$$\begin{cases} \partial_t u + \partial_x f(u) = 0, & t \in (0, T), \ x > 0, \\ u(0, x) = u_0(x), & x > 0, \\ u(t, 0) = \omega(t), & t \in (0, T), \end{cases} \tag{2.18}$$

with initial condition $u_0 \in \mathbf{L}^1(\mathbb{R}^+) \cap \mathbf{L}^\infty(\mathbb{R}^+)$, and boundary control $\omega \in \mathbf{L}^\infty(0, T)$. The definition of solution to (2.18) is the following one; see also Appendix A.

Definition 1 A solution to (2.18) is a function $u \in \mathbf{L}^1(\mathcal{D})$ such that the following conditions hold.

1. For every $k \in \mathbb{R}$ and for every $\varphi \in \mathbf{C}_\mathbf{c}^1(\mathcal{D}; \mathbb{R}^+)$

$$\int_\Omega \left[|u - k| \, \partial_t \varphi + \text{sgn}(u - k)(f(u) - f(k)) \, \partial_x \varphi \right] \mathrm{d}x \, \mathrm{d}t \geq 0.$$

2. There exists a set $\mathcal{E} \subseteq (0, T)$ with zero measure such that, for every $x > 0$,

$$\lim_{t \to 0, t \notin \mathcal{E}} \int_0^x u(t, \xi) \, \mathrm{d}\xi = \int_0^x u_0(\xi) \, \mathrm{d}\xi.$$

3. There exists a set $\mathcal{F} \subseteq (0, +\infty)$ with zero measure and two functions $\Upsilon : \mathbb{R}^+ \to \mathbb{R}$ and $\mu : \mathbb{R}^+ \to \{-1, 0, 1\}$ such that, for a.e. $t \in (0, T)$,

$$\lim_{x \to 0^+, \, x \notin \mathcal{F}} \int_0^t f\left(u\left(s, x\right)\right) \mathrm{d}s = \int_0^t \Upsilon(s) \, \mathrm{d}s,$$

$$\lim_{x \to 0^+, \, x \notin \mathcal{F}} \operatorname{sgn} f'\left(u\left(t, x\right)\right) = \mu(t)$$

and, for a.e. $t \in (0, T)$,

$$\begin{cases} \Upsilon(t) = f\left(\omega(t)\right), & \text{if } \mu(t) \in \{0, 1\}, \\ \Upsilon(t) \geq f\left(\omega(t)\right), & \text{if } \mu(t) = -1. \end{cases}$$

In [212], LeFloch proved that there exists a semigroup of solutions for (2.18), which satisfy the requirements of Definition 1. Given a set $\mathcal{U} \subseteq \mathbf{L}^\infty(0, T)$, let us define the attainable set (2.17), denoted here by $\mathcal{A}(T, \mathcal{U}, u_0)$ since there is no boundary control at $b = +\infty$. When the initial datum u_0 is the null function, then the following characterization holds.

Theorem 5 *In the case* $\mathcal{U} = \mathbf{L}^\infty(0, T)$ *and* $u_0 \equiv 0$, *then the attainable set* $\mathcal{A}(T, \mathbf{L}^\infty(0, T), 0)$ *is composed by all the functions* $w \in \mathbf{BV}(0, +\infty)$ *satisfying, for every* $x > 0$, *the following conditions:*

1. *if* $w(x) \neq 0$, *then* $f'(w(x)) \geq \frac{x}{T}$;
2. *if* $w(x-) \neq 0$ *and* $w(y) = 0$ *for every* $y > x$, *then* $f'(w(x-)) > \frac{x}{T}$;
3. *The upper Dini derivative* $D^+ w(x) := \lim\sup_{h \to 0} \dfrac{w(x + h) - w(x)}{h}$ *satisfies*

$$D^+ w(x) \leq \frac{f'(w(x))}{x f''(w(x))}.$$

The proof of Theorem 5, based on the concept of backwards characteristics (see [99, Chapter 11] or [98]), was proposed by Ancona and Marson in [12].

When dealing with optimal control problems for (2.18), it is important that the attainable set is a compact subset of $\mathbf{L}^1(\mathbb{R}^+)$. To achieve such a property, one has to restrict the set of admissible controls $\mathcal{U}$, as in the following result.

Theorem 6 *Fix* $N \in \mathbb{N} \setminus \{0\}$, $J \subseteq \mathbb{R}^+$, *and define* $\bar{u}$ *as the unique point of minimum for the flux* f. *Assume that*

1. $G : \mathbb{R}^+ \hookrightarrow [\bar{u}, +\infty)$ *is a measurable and uniformly bounded multifunction with convex and closed values;*
2. *for every* $i \in \{1, \cdots, N\}$, $q_i : \mathbb{R}^+ \times \mathbb{R} \to \mathbb{R}$ *is a measurable map, convex with respect to the second variable;*
3. *for every* $i \in \{1, \cdots, N\}$, $g_i : \mathbb{R}^+ \to \mathbb{R}$ *is a measurable map.*

Define the control set

$$\mathcal{U} = \left\{ \omega \in \mathbf{L}^\infty (0, T) : \begin{array}{l} \omega(t) \in G(t) \text{ for a.e. } t, \\ \int_0^t q_i \left(s, f \left(\omega(s) \right) \right) ds \leq g_i(t) \, \forall t \in J, \, \forall i \in \{1, \cdots, N\} \end{array} \right\}.$$

Then the attainable set $\mathcal{A}(T, \mathcal{U}, 0)$ is a compact subsets of $\mathbf{L}^1(\mathbb{R}^+)$ with respect the strong topology of $\mathbf{L}^1(\mathbb{R}^+)$.

For a proof see [12].

2.3.2   The Burgers' Equation with Two Controls

In this section, we consider the inviscid Burgers' equation

$$\partial_t u + \partial_x \left(\frac{u^2}{2} \right) = 0$$

on the domain $\mathcal{D} = (0, T) \times (a, b)$, with $T > 0$ fixed, $n = 1$, $\Omega = \mathbb{R}$, and $a < b$. The initial-boundary value problem in this situation reads

$$\begin{cases} \partial_t u + \partial_x f(u) = 0, & t \in (0, T), \, x \in (a, b), \\ u(0, x) = u_0(x), & x \in (a, b), \\ u(t, a) = \omega_a(t), & t \in (0, T), \\ u(t, b) = \omega_b(t), & t \in (0, T), \end{cases} \tag{2.19}$$

with initial condition $u_0 \in \mathbf{BV}(a, b)$, and boundary controls $\omega_a, \omega_b \in \mathbf{L}^\infty(0, T)$. For the definition of solution to (2.19) see Appendix A. For a later use, we define the set $\mathcal{B}$ composed by all the functions $w \in \mathbf{BV}([a, b])$ satisfying the following conditions:

1. $w(x-) \geq w(x+)$ for every $x \in (a, b)$;
2. the set $\{x \in (a, b) : w(x-) > w(x+)\}$ is at most countable;
3. for every $\bar{x} \in (a, b)$ such that $w(\bar{x}-) > 0$, then $w(x) > \frac{x-a}{T}$ for every $x \in (a, \bar{x})$;
4. for every $\bar{x} \in (a, b)$ such that $w(\bar{x}+) < 0$, then $w(x) < \frac{x-b}{T}$ for every $x \in (\bar{x}, b)$;
5. there exists at most one $\bar{x} \in (a, b)$ such that $w(x) > 0$ for every $x \in (a, \bar{x})$ and $w(x) < 0$ for every $x \in (\bar{x}, b)$;
6. for every $x \in (a, b)$, point of continuity for w such that $w(x) \neq 0$, the upper Dini derivative $D^+ w(x) := \limsup_{h \to 0} \dfrac{w(x + h) - w(x)}{h}$ satisfies

$$D^+ w(x) \leq \frac{b - a}{T}.$$

When the initial datum u_0 is the null function and the final time $T > 2(b - a)$, then the following characterization holds; see [179, Theorem 2.1].

Theorem 7 *In the case* $\mathcal{U}_a = \mathcal{U}_b = \mathbf{L}^\infty(0, T)$, $T > 2(b - a)$, *and* $u_0 \equiv 0$, *then the attainable set* $\mathcal{A}(T, \mathcal{U}_a, \mathcal{U}_b, 0)$ *contains the set* $\mathcal{B}$.

The proof is based on the *return method* introduced by Coron in [84].

In the case of a general initial datum u_0, then the following theorem holds; see [179, Theorem 1.1].

Theorem 8 *Assume* $\mathcal{U}_a = \mathcal{U}_b = \mathbf{L}^\infty(0, +\infty)$, $T > 2(b - a)$, *and* $u_0 \in$ **BV** $([a, b])$. *Then there exists* $T_c \geq T$, *called the time of approximate controllability, such that the attainable set* $\mathcal{A}(T_c, \mathcal{U}_a, \mathcal{U}_b, u_0)$ *contains the closure in the* $\mathbf{L}^1$-*topology of* $\mathcal{B}$.

2.3.3 Temple Systems on a Bounded Interval

In this section, we consider the system of conservation law (2.1) on the domain $\mathcal{D} = (0, T) \times (a, b)$, with $T > 0$ fixed, $n > 1$, $\Omega \subseteq \mathbb{R}^n$, and $a < b$. We assume that the system (2.1) is a strictly hyperbolic system of Temple type; see [265]. The initial-boundary value problem in this situation for (2.1) reads

$$\begin{cases} \partial_t u + \partial_x f(u) = 0, & t \in (0, T), \ x \in (a, b), \\ u(0, x) = u_0(x), & x \in (a, b), \\ u(t, a) = \omega_a(t), & t \in (0, T), \\ u(t, b) = \omega_b(t), & t \in (0, T), \end{cases} \tag{2.20}$$

with initial condition $u_0 \in \mathbf{L}^1(a, b)$, and boundary controls $\omega_a, \omega_b \in \mathbf{L}^\infty(0, T)$.

Before stating the main result, we need to introduce some notation and assumption. With $\lambda_1, \ldots, \lambda_n$ we denote the eigenvalues of the Jacobian matrix Df of the flux; see Appendix A.1 and A.6.2. The strictly hyperbolicity assumption implies that

$$\lambda_i(u_1) < \lambda_j(u_2)$$

for every $u_1, u_2 \in \Omega$ and $i, j \in \{1, \cdots, n\}$ with $i < j$. Moreover with $z_1, \cdots, z_n$ we denote a set of Riemann coordinates, so that the notation $z_i(u)$ stands for the i-th Riemann coordinate evaluated at the point u; see Appendix A.

Given, for every $i \in \{1, \cdots, n\}$, the real numbers $\alpha_i < \beta_i$, define the compact subset of Ω

$$\Gamma = \{u \in \Omega : z_i(u) \in [\alpha_i, \beta_i], i \in \{1, \cdots, n\}\}.$$

In the present setting, we suppose that the admissible controls are

$$\mathcal{U}_a = \mathcal{U}_b = \mathbf{L}^\infty((0, T); \Gamma)$$

and that the boundary is non-characteristic:

(NC)   there exist $p \in \{1, \cdots, n\}$ and $\lambda^{\min} > 0$ such that

$$\lambda_p(u) \leq -\lambda^{\min} < \lambda^{\min} \leq \lambda_{p+1}(u)$$

for every $u \in \Omega$.

For every $r > 0$, define the following sets

$$K_a^r = \left\{ \varphi \in \mathbf{L}^\infty\left((a,b); \Gamma\right) : \frac{z_i\left(\varphi(y)\right) - z_i\left(\varphi(x)\right)}{y - x} \leq \frac{r}{x - a}, \begin{array}{l} \text{a.e. } a < x < y < b \\ i \in \{p+1, \cdots, n\} \end{array} \right\}$$

$$K_b^r = \left\{ \varphi \in \mathbf{L}^\infty\left((a,b); \Gamma\right) : \frac{z_i\left(\varphi(y)\right) - z_i\left(\varphi(x)\right)}{y - x} \leq \frac{r}{b - y}, \begin{array}{l} \text{a.e. } a < x < y < b \\ i \in \{1, \cdots, p\} \end{array} \right\}$$

and

$$K^r = K_a^r \cap K_b^r. \tag{2.21}$$

The following result holds; see [11, Theorem 2.4 and Theorem 2.7].

Theorem 9 *Consider the system* (2.20) *with initial condition* $u_0 \in \mathbf{L}^1\left((a,b), \Gamma\right)$ *and boundary controls* $\omega_a \in \mathcal{U}_a$, $\omega_b \in \mathcal{U}_b$. *Assume that* (2.20) *is a strictly hyperbolic Temple system where each characteristic field is genuinely nonlinear and the non-characteristic condition* **(NC)** *holds.*
    Then:

1. *the attainable set* $\mathcal{A}(T, \mathcal{U}_a, \mathcal{U}_b, u_0)$ *is a compact subset of* $\mathbf{L}^1\left((a,b); \Gamma\right)$, *with respect to the strong topology;*
2. *for every* $\tau \in (0, T)$ *there exists* $r > 0$ *such that*

$$\mathcal{A}(t, \mathcal{U}_a, \mathcal{U}_b, u_0) \subseteq K^r$$

    for every $t \geq \tau$;
3. *if* $T > \frac{4(b-a)}{\lambda^{\min}}$, *then there exists* $r > 0$ *such that*

$$K^r \subseteq \mathcal{A}(T, \mathcal{U}_a, \mathcal{U}_b, u_0).$$

2.3.4 General Systems on a Bounded Interval

In this section, we consider the system of conservation law (2.1) on the domain $\mathcal{D} = (0, +\infty) \times (a,b)$, with $n > 1$, $\Omega \subseteq \mathbb{R}^n$, and $a < b$. We assume that the system (2.1) is a strictly hyperbolic system. The initial-boundary value problem in this situation reads

$$\begin{cases} \partial_t u + \partial_x f(u) = 0, & t > 0,\ x \in (a, b), \\ u(0, x) = u_0(x), & x \in (a, b), \\ u(t, a) = \omega_a(t), & t > 0, \\ u(t, b) = \omega_b(t), & t > 0, \end{cases} \tag{2.22}$$

with initial condition $u_0 \in \mathbf{L}^1 (a, b)$, and boundary controls $\omega_a, \omega_b \in \mathbf{L}^\infty (0, +\infty)$. We assume, similar to Sect. 2.3.3, that the system is strictly hyperbolic, that each characteristic field is either genuinely nonlinear or linearly degenerate, and that the non-characteristic condition **(NC)** holds. Few results are available in the present setting. In particular, in this part we state a positive result, dealing with asymptotic stabilization, a negative result about the local controllability around a constant state, and a positive result for the p-system.

Theorem 10 ([52, Theorem 1]) *Fix K, a compact and connected subset of Ω. There exist positive constants C_0, δ, and κ such that, for every $u^* \in K$, for every initial datum $u_0 \in \mathbf{L}^1 ((a, b); K)$ with $\mathrm{TV}(u_0) < \delta$, and for every $t > 0$,*

$$\inf \left\{ \| u - u^* \|_{\mathbf{L}^\infty} : u \in \mathcal{A}(t, u_0, \mathcal{U}_a, \mathcal{U}_b),\ \mathrm{TV}(u) \le C_0 e^{-2\kappa t} \right\} \le C_0 e^{-2\kappa t}.$$

The following result gives a negative answer about the exact controllability around constant states for a class of 2×2 systems, satisfying the following conditions:

$$\begin{cases} \nabla\lambda_1(u) \cdot r_1(u) > 0 \\ \nabla\lambda_2(u) \cdot r_2(u) > 0 \end{cases} \qquad \forall u \in \Omega \tag{2.23}$$

and

$$\begin{cases} r_1(u) \wedge r_2(u) < 0 \\ r_1(u) \wedge (Dr_1(u)r_1(u)) < 0 \qquad \forall u \in \Omega, \\ r_2(u) \wedge (Dr_2(u)r_2(u)) < 0 \end{cases} \tag{2.24}$$

where r_1 and r_2 form a basis of right eigenvectors of the Jacobian matrix Df (see Appendix A.6.2) and $\wedge$ denotes the wedge product, i.e., if $v = (v_1, v_2)$ and $w = (w_1, w_2)$, then $v \wedge w = v_1 w_2 - v_2 w_1$. Note that condition (2.24) implies that the interaction of two shocks of the same family generates a shock in the other family.

Theorem 11 ([52, Theorem 2]) *Fix $n = 2$. Assume that (2.22) is a strictly hyperbolic system, genuinely nonlinear, satisfying the non-characteristic condition **(NC)** with $p = 1$, (2.23) and (2.24).*

Then, for every $\varepsilon > 0$, there exists an initial datum $u_0 \in \mathbf{L}^1 (a, b)$ with $\mathrm{TV}(u_0) \le \varepsilon$ such that, for every $t > 0$, all the elements in $\mathcal{A}(t, u_0, \omega_a, \omega_b)$ have a countable *number of shocks. In particular, the attainable set $\mathcal{A}\,(t,u_0,\omega_a,\omega_b)$ does not contain constant states.*

In the case condition (2.24) does not hold, there exist some systems, where constant states can be reachable in finite time. For example, the p-system in Eulerian or Lagrangian coordinates has such property; see [149, 150]. Indeed consider the system

$$\begin{cases} \partial_t \rho + \partial_x q = 0, & t > 0,\ x \in (a,b), \\ \partial_t q + \partial_x \left(\frac{q^2}{\rho} + \kappa \rho^\gamma \right) = 0, & t > 0,\ x \in (a,b), \\ (\rho, q)\,(0, x) = (\rho_0(x), q_0(x)), & x \in (a,b), \\ (\rho, q)\,(t, a) = \omega_a(t), & t > 0, \\ (\rho, q)\,(t, b) = \omega_b(t), & t > 0, \end{cases} \tag{2.25}$$

where $\kappa > 0$ and $1 < \gamma \le 3$. For later use, we set

$$c_\gamma = \frac{1}{\frac{1}{2} + \frac{\gamma - 1}{4\sqrt{\gamma}}} \tag{2.26}$$

and we denote by w^1 and w^2 the pairs of Riemann invariants, i.e.,

$$w^1\,(\rho, q) = \frac{q}{\rho} + \frac{2\sqrt{\kappa\gamma}}{\gamma - 1}\rho^{\frac{\gamma-1}{2}} \qquad w^2\,(\rho, q) = \frac{q}{\rho} - \frac{2\sqrt{\kappa\gamma}}{\gamma - 1}\rho^{\frac{\gamma-1}{2}}. \tag{2.27}$$

The following controllability result holds.

Theorem 12 ([149, Theorem 1]) *Let $(\bar{\rho}_0, \bar{q}_0)$, $(\bar{\rho}_1, \bar{q}_1)$ be constant states in $(0, +\infty) \times \mathbb{R}$. Set $\bar{\lambda}_1 = \lambda_1\,(\bar{\rho}_1, \bar{q}_1)$ and $\bar{\lambda}_2 = \lambda_2\,(\bar{\rho}_1, \bar{q}_1)$. Then there exist $\varepsilon_1 > 0$, ε_2, and $T > 0$, such that for every (ρ_0, q_0), $(\rho_1, q_1) \in \mathbf{BV}\,([a, b]; (0, +\infty) \times \mathbb{R})$ satisfying*

$$\|\rho_0 - \bar{\rho}_0\|_{\mathbf{L}^\infty(a,b)} + \|q_0 - \bar{q}_0\|_{\mathbf{L}^\infty(a,b)} \le \varepsilon_1 \qquad \mathrm{TV}\,(\rho_0, q_0) \le \varepsilon_1,$$

$$\|\rho_1 - \bar{\rho}_1\|_{\mathbf{L}^\infty(a,b)} + \|q_1 - \bar{q}_1\|_{\mathbf{L}^\infty(a,b)} \le \varepsilon_2 \qquad \mathrm{TV}\,(\rho_1, q_1) \le \varepsilon_2,$$

and, for every $a \le x < y \le b$,

$$\frac{w^2\,(\rho_1(x), q_1(x)) - w^2\,(\rho_1(y), q_1(y))}{x - y} \le \frac{c_\gamma}{2} \max\left\{ \frac{\bar{\lambda}_2 - \bar{\lambda}_1}{1 - y}, \frac{\bar{\lambda}_1}{x}, -\frac{\bar{\lambda}_1}{1 - y} \right\}$$

$$\frac{w^1\,(\rho_1(x), q_1(x)) - w^1\,(\rho_1(y), q_1(y))}{x - y} \le \frac{c_\gamma}{2} \max\left\{ \frac{\bar{\lambda}_2 - \bar{\lambda}_1}{x}, -\frac{\bar{\lambda}_2}{1 - y}, \frac{\bar{\lambda}_2}{x} \right\},$$

where c_γ, w^1, and w^2 are defined in (2.26)–(2.27), there is a weak entropy admissible solution (ρ, q) (see Definition 47 in Appendix A.8) to (2.25) in $[0, T] \times$

$[a, b]$ such that $(\rho, q)(0, x) = (\rho_0, q_0)(x)$ and $(\rho, q)(T, x) = (\rho_1, q_1)(x)$ for a.e. $x \in [a, b]$.

2.4 Lyapunov Stabilization of Scalar Conservation Laws with Two Boundaries

In this section, we consider stabilization problems for the scalar conservation law (2.1) on the domain $\mathcal{D} = (0, T) \times (a, b)$, with $T > 0$ fixed, $n = 1$, $\Omega = \mathbb{R}$, and $a < b$, which reads

$$\begin{cases} \partial_t u + \partial_x f(u) = 0, & t \in (0, T),\ x \in (a, b), \\ u(0, x) = u_0(x), & x \in (a, b), \\ u(t, a) = \omega_a(t), & t \in (0, T), \\ u(t, b) = \omega_b(t), & t \in (0, T), \end{cases} \qquad (2.28)$$

with initial condition $u_0 \in \mathbf{L}^1(a, b)$, and boundary data $\omega_a, \omega_b \in \mathbf{L}^\infty(0, T)$. In this section, we assume that the flux f is a strictly convex smooth function such that

$$\lim_{u \to \pm\infty} f(u) = +\infty.$$

The case of concave flux function is entirely similar. We denote with u_m the unique point of minimum for f.

The interest is stemming out from many applications, in particular from traffic flow, where the interval $[a, b]$ represent a stretch of road and the controls are possible only at the boundary points, e.g., via controlled access, ramp metering, or traffic lights.

2.4.1 Approximation of Solutions via Piecewise Smooth Functions

Here we approximate **BV** solutions to (2.28) using a special class of piecewise smooth functions, denoted by **PWS**$^+$. In the next subsections, Lyapunov stability analysis for solutions to (2.28) is performed for functions in **PWS**$^+$.

Definition 2 We define **PWS**$^+$ as the class of PieceWise Smooth functions $v : (a, b) \to \mathbb{R}$ such that there exist a finite number of points $a = x_0 < x_1 < \cdots < x_N = b$ (depending on v) such that

1. v is bounded and of class $\mathbf{C}^\infty$ on the intervals (x_{j-1}, x_j) for $j \in \{1, \ldots, N\}$;
2. $v'(x) \geq 0$ for every $x \in (x_{j-1}, x_j)$ and $j \in \{1, \ldots, N\}$;
3. v has only downward jumps, i.e. $v(x_j-) \geq v(x_j+)$ for $j \in \{1, \ldots, N - 1\}$.

We note that every **BV** solution to (2.28) can be approximated, in the $\mathbf{L}^1$ topology, by a function in the class **PWS$^+$**.

Theorem 13 *Consider $T > 0$, $a < b$, and let $u_0 : (a, b) \mapsto \mathbb{R}$, $\omega_a, \omega_b : (0, T) \mapsto \mathbb{R}$ be functions with bounded total variation. Then, for every $\varepsilon > 0$, there exist $u^\varepsilon \in \mathbf{PWS}^+$ and piecewise constant boundary data $\omega_a^\varepsilon, \omega_b^\varepsilon : (0, T) \mapsto \mathbb{R}$ with*

$$\left\| \omega_a - \omega_a^\varepsilon \right\|_{\mathbf{L}^1(0,T)} \leq \varepsilon \qquad \left\| \omega_b - \omega_b^\varepsilon \right\|_{\mathbf{L}^1(0,T)} \leq \varepsilon$$

such that the solutions u and u^ε to (2.28) respectively with initial-boundary data $(u_0, \omega_a, \omega_b)$ and with $\left(u_0^\varepsilon, \omega_a^\varepsilon, \omega_b^\varepsilon \right)$ satisfy, for every $0 \leq t \leq T$,

$$\left\| u(t) - u^\varepsilon(t) \right\|_{\mathbf{L}^1(a,b)} \leq \varepsilon.$$

It is also interested to note that solutions to (2.28) with initial data in **PWS$^+$** remain in that class, provided the boundary conditions are piecewise constant.

Theorem 14 *Fix $T, \delta > 0$, $a < b$, and let $u_0 \in \mathbf{PWS}^+$ and $\omega_a, \omega_b : (0, T) \mapsto \mathbb{R}$ be piecewise constant. Then the solution u to (2.28) satisfies $u(t) \in \mathbf{PWS}^+$ for all times $0 \leq t \leq T$.*

The proofs of Theorem 13 and of Theorem 14 can be found in [40, Theorem 2 and Theorem 3].

2.4.2 Lyapunov Functional

Here we introduce a Lyapunov functional to stabilize (2.28) within the class **PWS$^+$**. We fix a constant state $u^* \in \Omega$ and for every solution u to (2.28), we consider its perturbation around the steady state u^*; thus define $\tilde{u} = u - u^*$. The aim is to stabilize the solution to u^*.

Since the results in Sect. 2.4.1, we assume that u is in **PWS$^+$** and consider the classical Lyapunov function candidate [195, 197]:

$$V(t) = \frac{1}{2} \int_a^b \tilde{u}^2(t, x) \, \mathrm{d}x = \frac{1}{2} \int_a^b (u(t, x) - u^*)^2 \, \mathrm{d}x. \tag{2.29}$$

Notice that $t \mapsto u(t, \cdot)$ is continuous from $[0, T]$ to $\mathbf{L}^1(a, b)$, and the function $V(\cdot)$ is well defined and continuous. We index the jump discontinuities of $u(t, \cdot)$ in increasing order of their locations at time t by $i = 0, \ldots, N(t)$, including for notational purposes the boundaries a, b, with $x_0(t) = a$ and $x_N(t) = b$, and write:

$$V(t) = \frac{1}{2} \sum_{i=0}^{N(t)-1} \int_{x_i(t)}^{x_{i+1}(t)} \tilde{u}^2(t, x) \, \mathrm{d}x. \tag{2.30}$$

From Theorem 14, we know that for all integers $i = 0, \dots, N(t)$, the function $u(t, \cdot)$ is smooth in the domain $(x_i(t), x_{i+1}(t))$. Moreover, the trajectories $x_i(\cdot)$ are differentiable with speed given by the Rankine–Hugoniot relation; see (A.18). Therefore the function $V(\cdot)$ is differentiable at any time except interaction times of discontinuities, which is a finite set. More precisely, for every time t such that $N(t)$ is locally constant and traces are continuous, differentiating expression (2.30) we get:

$$\frac{dV}{dt}(t) = \frac{1}{2} \sum_{i=0}^{N(t)-1} \int_{x_i(t)}^{x_{i+1}(t)} \partial_t \tilde{u}^2 \, dx$$

$$+ \frac{1}{2} \sum_{i=0}^{N(t)-1} \left[\tilde{u}^2(t, x_{i+1}(t)-) \frac{dx_{i+1}}{dt}(t) - \tilde{u}^2(t, x_i(t)+) \frac{dx_i}{dt}(t) \right].$$

We can write $\partial_t \tilde{u}^2 = 2\tilde{u} \, \partial_t \tilde{u}$ and $\partial_t \tilde{u} = -\partial_x f(\tilde{u} + u^*)$. Integrating by parts and indicating by F a primitive of the flux f, we get:

$$\frac{dV}{dt}(t) = \tilde{u}(t, a) f(\tilde{u}(t, a) + u^*) - \tilde{u}(t, b) f(\tilde{u}(t, b) + u^*)$$

$$- F(\tilde{u}(t, a) + u^*) + F(\tilde{u}(t, b) + u^*)$$

$$+ \sum_{i=1}^{N(t)-1} \Delta_i \left(\tilde{u} f(\tilde{u} + u^*) - F(\tilde{u} + u^*) \right) \tag{2.31}$$

$$- \sum_{i=1}^{N(t)-1} \frac{\tilde{u}(t, x_i-) + \tilde{u}(t, x_i+)}{2} \Delta_i f(\tilde{u} + u^*),$$

where Δ_i gives the jump at the i-th discontinuity and we used the Rankine–Hugoniot relation to express the speed of i-th discontinuity. Notice that the first four terms depend on the boundary trace of the solution, while the last two terms depend on the shock dynamics inside the domain. Therefore last two terms are not controllable with boundary control, but they have a stabilizing effect on the Lyapunov function:

Proposition 1 *Given a fixed state $u^* \in \Omega$ and a solution u to (2.28), then the following inequality holds:*

$$\sum_{i=1}^{N(t)-1} \left[\Delta_i \left(\tilde{u} f(\tilde{u} + u^*) - F(\tilde{u} + u^*) \right) - \frac{\tilde{u}(t, x_i-) + \tilde{u}(t, x_i+)}{2} \Delta_i f(\tilde{u} + u^*) \right] \le 0,$$

which implies that the jump discontinuity dynamics contributes to the decrease of the Lyapunov function (2.29).

For a proof, see [40, Proposition 1].

Remark 1 Stability of the jump discontinuity dynamics is implied by the Oleinik entropy condition. Thus, for a convex flux, the internal dynamics is strictly stabilizing, i.e., we have a strict decrease of the Lyapunov function.

Remark 2 Possibly except at discontinuity interaction times, the internal dynamics is stabilizing letting the Lyapunov function (2.29) decay. This is critical for boundary stabilization where the control action has no effect inside the domain.

Remark 3 At a time t at which the number of discontinuities is not constant or the boundary trace is not continuous, the Lyapunov function is not differentiable, however, the difference between the right and left derivative at t^+ and t^-, respectively, can be computed. This is addressed in Sect. 2.4.4.

2.4.3   Control Space and Lyapunov Stability

In this part we introduce control spaces and we show the existence of a boundary control such that the functional V in (2.29) is decreasing and the system is Lyapunov stable. We first define the control spaces at $x = a$ and at $x = b$.

Definition 3 The *control space* $\mathcal{C}_a$ (resp. $\mathcal{C}_b$) is composed by all the couples (u_l, u_r) such that the Riemann problem

$$
\begin{cases}
\partial_t u + \partial_x f(u) = 0 \\
u(0, x) = \begin{cases} u_l, & x < 0, \\ u_r, & x > 0, \end{cases}
\end{cases}
$$

is solved with waves with non-negative (resp. non positive) speed.

The control spaces $\mathcal{C}_a$ and $\mathcal{C}_b$ can be characterized in the following way; for a proof see [40, Proposition 2].

Proposition 2 *Let* u_m *be the unique point of minimum for the flux function* f. *A couple* (u_l, u_r) *belongs to* $\mathcal{C}_a$ *if and only if*

$$
\begin{aligned}
\text{either} \quad & u_l = u_r \\
\text{or} \quad & u_l \geq u_m, \quad u_r \geq u_m \\
\text{or} \quad & u_l \geq u_m, \quad u_r \leq u_m, \quad f(u_l) > f(u_r).
\end{aligned}
\tag{2.32}
$$

A couple (u_l, u_r) *belongs to* $\mathcal{C}_b$ *if and only if*

$$
\begin{aligned}
\text{either} \quad & u_l = u_r \\
\text{or} \quad & u_l \leq m, \quad u_r \leq m \\
\text{or} \quad & u_l \geq m, \quad u_r \leq m, \quad f(u_l) < f(u_r).
\end{aligned}
\tag{2.33}
$$

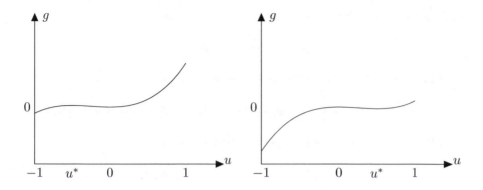

Fig. 2.1 The graph of the function g defined in (2.34), in the case of a Burgers flux function $f(u) = u^2$ and $F(u) = \frac{u^3}{3}$, so that $u_m = 0$. At the left, the case $u^* = -\frac{1}{2}$, while, at the right, the case $u^* = \frac{1}{2}$

The following stability result holds.

Theorem 15 *There exist boundary conditions ω_a, ω_b such that, if the solution u to the initial-boundary value problem (2.28) is in the class* **PWS**$^+$, *then*

1. *for a.e. $t \in [0, T]$, $(\omega_a(t), u(t, a+)) \in C_a$;*
2. *for a.e. $t \in [0, T]$, $(u(t, b-), \omega_b(t)) \in C_b$;*
3. *the functional $t \mapsto V(t)$, defined in (2.29), is strictly decreasing;*
4. *the system (2.28) is Lyapunov stable.*

For a proof see [40, Theorem 4]. It is possible to describe explicitly a class of controls described in Theorem 15. To this aim, we first consider a function

$$
\begin{aligned}
g : \mathbb{R} &\longrightarrow \mathbb{R} \\
u &\longmapsto (u - u^*) \, f(u) - F(u),
\end{aligned}
\tag{2.34}
$$

where F is a primitive of the flux f; see Fig. 2.1.

Lemma 1 *Let u_m denote the unique point of minimum of f. The smooth function g satisfies the following properties.*

1. *g is strictly increasing in $(-\infty, \min\{m, u^*\}) \cup (\max\{m, u^*\}, +\infty)$ and strictly decreasing in $(\min\{m, u^*\}, \max\{m, u^*\})$.*
2. *For $u > v$ such that $f(u) = f(v)$, we have $g(u) > g(v)$.*

The set of boundary controls $(\omega_a(t), \omega_b(t))$ of Theorem 15, which stabilize the system (2.28), in the case $u^* < u_m$, can be chosen according to the following cases.

1. If $(u(t, a+), u(t, b-)) \in [m, +\infty) \times (m, +\infty)$, then

$$\omega_a(t) \in [m, u(t, a+)) \qquad \omega_b(t) = u(t, b-).$$

2. If $(u(t, a+), u(t, b-)) \in [m, +\infty) \times (-\infty, m]$, then

$$\omega_a(t) \in (m, +\infty) \qquad \omega_b(t) \in (-\infty, m) \qquad g(\omega_a(t)) = g(\omega_b(t)).$$

3. If $(u(t, a+), u(t, b-)) \in (-\infty, m) \times (-\infty, m]$, then

$$\omega_a(t) = u(t, a+) \qquad \omega_b(t) = u^*.$$

4. If $(u(t, a+), u(t, b-)) \in (-\infty, m) \times (m, +\infty]$ and $g(u(t, a+)) < g(u(t, b-))$, then

$$\omega_a(t) = u(t, a+) \qquad \omega_b(t) = u(t, b-).$$

5. If $(u(t, a+), u(t, b-)) \in (-\infty, m) \times (m, +\infty]$ and $g(u(t, a+)) \geq g(u(t, b-))$, then

$$\omega_a(t) = u(t, a+) \qquad \omega_b(t) = u^*.$$

Remark 4 The boundary controls in Theorem 15 provide Lyapunov stability, but not asymptotic stability in general. In the following section, a greedy controller is defined and it maximizes the instantaneous decrease of the Lyapunov function (but also not guaranteeing asymptotic stability). Finally, in Sect. 2.4.5 we design an improved controller which guarantees asymptotic stability and invariance of the class **PWS**$^+$.

2.4.4 Greedy Controls

Here we characterize the values of boundary controls, that minimize the derivative of the Lyapunov function V (2.29). Since the type of waves generated at the boundaries influences the value of the derivative of V, we first provide a detailed description of them. Table 2.1 summarizes the types of waves created at the left boundary for controls taking values in the control space $\mathcal{C}_a$; see Definition 3.

Let us characterize the changes in the Lyapunov function, due to the variation in the number of shock waves in the solution, resulting both from internal interactions and from the generation or absorption of waves at the boundaries.

Proposition 3 *Fix a time $\bar{t}$ at which the number of jump discontinuities changes.*

1. *If two shock waves interact at time $\bar{t}$, then the derivative of the Lyapunov function V decreases, i.e., $V'(\bar{t}-) > V'(\bar{t}+)$.*

Table 2.1 Waves exiting from the left boundary depending from the boundary control and the trace of the solution at this boundary

	$u(t, a+) < u_m$	$u(t, a+) \geq u_m$
$\omega_a \geq u_m$	$f(\omega_a) > f(u(t, a+))$: Shock	$\omega_a > u(t, a+)$: Shock
		$\omega_a = u(t, a+)$: No wave
		$\omega_a < u(t, a+)$: Rar. wave
$\omega_a < u_m$	$\omega_a = u(t, a+)$: No wave	Rarefaction with vanishing boundary trace

2. *Assume that a discontinuity wave crosses the left boundary and let us denote u^- the value of the boundary trace at time $\bar{t}-$ and u^+ the value of the boundary trace at time $\bar{t}+$. The jump in the derivative of the Lyapunov function reads*

$$V'(\bar{t}+) - V'(\bar{t}-) = \left(f(u^+) - f(u^-) \right) \left(\frac{u^- + u^+}{2} - u^* \right), \qquad (2.35)$$

where the term $\left(f(u^+) - f(u^-) \right)$ is positive, since f is a convex flux. Moreover we have the following cases.

(a) *If $u^* < \frac{u^- + u^+}{2}$, then there is an increase in the derivative of the Lyapunov function V.*

(b) *If $u^* > \frac{u^- + u^+}{2}$, then there is a decrease in the derivative of the Lyapunov function V.*

The case of right boundary can be treated similarly with a change in the sign in (2.35).

For a proof see [40, Proposition 3]. Proposition 3 can be restated in the case of a concave flux function. The next result selects the boundary controls that maximize the decrease rate of the Lyapunov function V in two different cases. In the first one only rarefaction waves are created, while in the second one shock waves are produced.

Proposition 4 ([40, Proposition 4]) *Let u be a solution to the initial-boundary value problem (2.28) and let g be the function introduced in (2.34).*

1. *The boundary controls $t \mapsto \omega_a^r$ and $t \mapsto \omega_b^r$ that minimize the decrease of the Lyapunov function V without introducing shock waves are given by:*

$$\omega_a^r(t) \doteq \arg\min \{ g(u) : (u, u(t, a+)) \in \mathcal{C}_a, \ u \leq u(t, a+) \}$$

$$\omega_b^r(t) \doteq \arg\max \{ g(u) : (u, u(t, b-)) \in \mathcal{C}_b, \ u \geq u(t, b-) \}.$$

2. *The boundary controls $t \mapsto \omega_a^s(t)$ and $t \mapsto \omega_b^s(t)$ that minimize the decrease of the Lyapunov function V by introducing discontinuities at the boundaries, can be obtained by:*

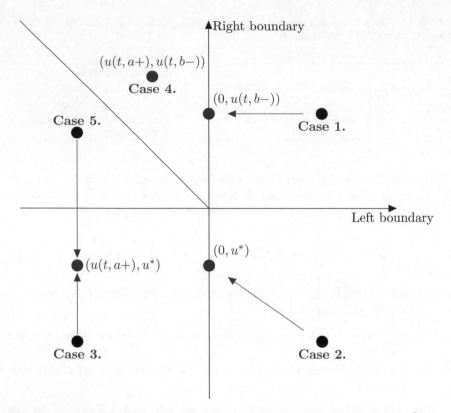

Fig. 2.2 Construction of the greedy stabilizing controller for the Burgers flux $f(u) = u^2$ in the case $u^* < 0 = u_m$. The cases correspond to the ones of the procedure described in Sect. 2.4.4

$$\omega_a^s(t) \doteq \arg\min\{S(u(t,a+),u) : (u,u(t,a+)) \in \mathcal{C}_a,\ u > u(t,a+)\}$$
$$\omega_b^s(t) \doteq \arg\max\{S(u(t,b-),u) : (u,u(t,b-)) \in \mathcal{C}_b,\ u < u(t,b-)\},$$

where

$$S(u,v) := (f(v) - f(u))\left(\frac{u+v}{2} - u^*\right).$$

In the case $u^* < u_m$, a greedy boundary control $(\omega_a(t), \omega_b(t))$, maximizing the instantaneous decrease of the Lyapunov function (2.29), can be constructed in the following way; see Fig. 2.2.

Case 1.   If $(u(t,a+), u(t,b-)) \in [u_m, +\infty) \times (u_m, +\infty)$, then

$$\omega_a(t) = u_m, \qquad \omega_b(t) = u(t,b-).$$

Case 2.   If $(u(t,a+), u(t,b-)) \in [u_m, +\infty) \times (-\infty, u_m]$, then

$$\omega_a(t) = u_m, \qquad \omega_b(t) = u^*.$$

Case 3. If $(u(t, a+), u(t, b-)) \in (-\infty, u_m) \times (-\infty, u_m]$, then

$$\omega_a(t) = u(t, a+), \qquad \omega_b(t) = u^*.$$

Case 4. If $(u(t, a+), u(t, b-)) \in (-\infty, u_m) \times (u_m, +\infty)$ and $g(u(t, b-)) > g(u^*)$, then

$$\omega_a(t) = u(t, a+), \qquad \omega_b(t) = u(t, b-).$$

Case 5. If $(u(t, a+), u(t, b-)) \in (-\infty, u_m) \times (u_m, +\infty)$ and $g(u(t, b-)) \le g(u^*)$, then

$$\omega_a(t) = u(t, a+), \qquad \omega_b(t) = u^*.$$

We note in Example 1 that with the greedy controller constructed in this part asymptotic stability may not be obtained. We also illustrate that the naive *brute force* control $(\omega_a(t) = u^*, \omega_b(t) = u^*)$ may create oscillations at the boundary.

Example 1 Choose $u^* < u_m$ and define $\hat{u}$ by $u_m < \hat{u}$ and $f(\hat{u}) = f(u^*)$. Given $0 < \Delta < (a+b)/2$, such that $\frac{b-a}{4\Delta} \in \mathbb{N}$, and $0 < k < \hat{u}$ we consider the following initial datum

$$u_0(x) = \begin{cases} u_m, & x \in \left(u, \frac{a+b}{2}\right) \\ \hat{u} - k, & x \in \left(\frac{a+b}{2} + (2\,p)\,\Delta, \frac{a+b}{2} + (2\,p+1)\,\Delta\right) \\ & p \in \left\{0, \cdots, \frac{b-a}{4\Delta} - 1\right\} \\ \hat{u} + k, & x \in \left(\frac{a+b}{2} + (2\,p+1)\,\Delta, \frac{a+b}{2} + (2\,p+2)\,\Delta\right) \\ & p \in \left\{0, \cdots, \frac{b-a}{4\Delta} - 1\right\}. \end{cases}$$

In such a case, **Case 1.** applies, hence the boundary controls are $(\omega_a(t) = u_m, \omega_b(t) = u(t, b-))$. The characteristic speed of u_m is zero, thus the right boundary value converges toward u_m in infinite time. The solution remains in the configuration characterized by **Case 1.**, and converges to the steady state u_m not reaching the target density u^*. This shows that stability is achieved but not asymptotic stability.

The brute force control $(\omega_a(t) = u^*, \omega_b(t) = u^*)$ has no action on the system when $u(t, b-) = \hat{u} + k$, since the control values are outside of the control space C_b. While when $u(t, b-) = \hat{u} - k$, the brute force control induces slow backward moving shock waves $(\hat{u} - k, u^*)$ from the right boundary which interact with fast forward moving shock waves $(\hat{u} + k, \hat{u} - k)$ coming from the initial datum, and create slow forward moving shock waves $(\hat{u} + k, u^*)$. Hence we observe large oscillations at the right boundary, comparable to the oscillations in the initial datum. More precisely, the trace at the boundary $x = b$ oscillates between the value u^* and values in the interval $[\hat{u}-k, \hat{u}+k]$, generating a total variation in time which satisfies

$$\mathrm{TV}_t(u(t, b-)) \geq \frac{b-a}{4\Delta}(\hat{u} - k - u^*).$$

Since Δ is arbitrary, the boundary oscillations can be arbitrarily large, but in any case proportional to the oscillations in the initial datum. Eventually all the waves generating by an oscillating initial datum exit the domain and the naive control produces backward moving shock waves (u_m, u^*) which yield convergence.

2.4.5 Lyapunov Asymptotic Stability

Here we design a control strategy, which guarantees Lyapunov asymptotic stability. More precisely, Theorem 16 shows that the associated solution to (2.28) remains in the class **PWS**$^+$ for initial data u_0 in the same class. Moreover, in Theorem 17 we state **BV** estimates for the solution and the boundary controls. Finally, in Theorem 18, we use the **BV** estimates to extend the construction to **BV** initial data and we state Lyapunov asymptotic stability.

In this part, we restrict to the case $u^* < u_m$ for simplicity. Define $\hat{u} > u_m$ by $f(\hat{u}) = f(u^*)$, $\check{u} > u_m$ by $g(\check{u}) = g(u^*)$, and $\bar{u} = \frac{\hat{u}+\check{u}}{2}$. By Lemma 1, we deduce that $\check{u} < \hat{u}$ and thus $\check{u} < \bar{u} < \hat{u}$. The specific choice of $\bar{u}$ guarantees the decrease of $V(\cdot)$ (as it would any other control in $]\check{u}, \hat{u}[.)$ The feedback controls $\omega_a(t)$, $\omega_b(t)$, are defined as:

$$\begin{cases} (u_m, u(t, b-)) & \text{if } (u(t, a+), u(t, b-)) \in [u_m, +\infty) \times (\bar{u}, +\infty), \\ (u_m, u^*) & \text{if } (u(t, a+), u(t, b-)) \in [u_m, +\infty) \times (-\infty, \bar{u}], \\ (u(t, a+), u^*) & \text{if } (u(t, a+), u(t, b-)) \in (-\infty, u_m) \times (-\infty, \bar{u}], \\ (u(t, a+), u(t, b-)) & \text{if } (u(t, a+), u(t, b-)) \in (-\infty, u_m) \times (\bar{u}, +\infty). \end{cases}$$
$$(2.36)$$

Notice that the boundary controls are of feedback type, since they depend only on the trace of the unknown at the boundaries. Instead, the "nonlocal" boundary controls, defined in Sect. 2.4.6, depend also on the initial state.
We have the following result.

Theorem 16 *Consider an initial datum u_0 in **PWS**$^+$ and the boundary controls given by formula (2.36). Then the corresponding boundary value problem admits a unique solution, which belongs to the class **PWS**$^+$ for all times. Moreover, the function $V(\cdot)$ is strictly decreasing along the solution and the equation is stable in the sense of Lyapunov.*

To extend the result to general initial data in **BV**, we need estimates on the total variation in time of the controls $\mathrm{TV}_t(\omega_a)$, $\mathrm{TV}_t(\omega_b)$, and in space of the generated solution $\mathrm{TV}_x(u(t, \cdot))$, which are given by the next result.

Theorem 17 *Consider an initial datum u_0 in* **PWS**$^+$*, the boundary controls given by Case 1.–Case 5., and let us indicate by $u(t, x)$ the corresponding solution. Then, defining $C = 2 \left(\sup_x |u_0(x) - u_m| + |u_m - u^*| \right)$, we have the following estimates:*

$$\mathrm{TV}_x(u(t, \cdot)) \leq \mathrm{TV}_x(u_0) + C + |\bar{u} - u^*| \tag{2.37}$$

$$\mathrm{TV}_t(\omega_a) \leq \mathrm{TV}_x(u_0) + C \tag{2.38}$$

$$\mathrm{TV}_t(\omega_b) \leq \mathrm{TV}_x(u_0) + C + |\bar{u} - u^*| \cdot \frac{\mathrm{TV}_x(u_0) + C}{|\bar{u} - \hat{u}|}. \tag{2.39}$$

The proof of Theorem 17 is based on careful estimates on the flux variation and possible wave patterns generated by the boundary controls; see [40, Theorem 6]. We are now ready to state the last result of this section [40, Theorem 7].

Theorem 18 *Consider an initial datum u_0 in* **BV** *and the boundary controls given by formula (2.36), then there exists a unique entropic solution to the corresponding initial-boundary problem such that (2.37), (2.38) and (2.39) hold true. Moreover, $\lim_{t \to +\infty} V(t) = 0$, i.e. $\lim_{t \to +\infty} \|u(t, \cdot) - u^*\|_{L^2} = 0$.*

To prove Theorem 18, one first uses standard compactness and Helly's Theorem, then observe that the solution attains the boundary value and, finally, use the decay to N-wave solutions, see [226]. Notice that, because of the BV estimates, the convergence of u actually holds in all L^p norms.

2.4.6 Nonlocal Controls

As we noticed the greedy control may not stabilize the system to u^*, while the brute force control $u_a \equiv u_b = u^*$ may overshoot and produce oscillations. Finally, control (2.36) stabilizes the system, but the stabilization time can be far from optimal. Therefore, in this section, we show *nonlocal* controls $\omega_a^{nl} \, \omega_b^{nl}$, which fast stabilize the system to u^*. We use the term nonlocal to indicate that these controls depend not only on the values of the traces $u(t, a+)$ and $u(t, b-)$.

We focus again, for simplicity, only on the case $u^* < u_m$. Define $A = \sup_{x \in [a,b]} u_0(x)$ and $\hat{A} < u_m$ be such that $f(\hat{A}) = f(A)$. For every $U < \hat{A}$ we define:

$$T_1(U) = \frac{(b-a)(A-U)}{f(U) - f(A)}, \quad T_0(U) = T_1 - \frac{(b-a)}{|f'(U)|} \tag{2.40}$$

and set ω_a^{nl} as in (2.36), while:

$$\omega_b^{nl}(t) = \begin{cases} U, & 0 \leq t \leq T_0, \\ u^*, & T_0 < t < +\infty. \end{cases} \tag{2.41}$$

The meaning of such construction is as follows. First, in the same spirit as [149, 150, 179], we send a large shock $(u_0(a+), U)$ with negative speed to move the system in the zone $u < u_m$ and then apply the stabilizing control. Notice that T_1 is computed as the maximal time taken by the big shock to cross the interval $[a, b]$, while T_0 is the time at which the characteristic corresponding to u^* should start from b to reach a at time T_1. These choices will guarantee the desired effect. Notice also that T_0 is a safe choice, but smaller values may give a better performance.

2.4.7 Numerical Examples

In this section, we present numerical results obtained for a benchmark scenario. The numerical scheme used here is the standard Godunov scheme [154] with 200 cells in space and a time discretization satisfying the tight *Courant-Friedrich-Levy* (CFL) condition [216]. We consider the flux function $u \mapsto u^2/2$, the equilibrium state $u^* = -1$, and the space domain $[0, 1]$ with the oscillating initial condition

$$u_0(x) = 1 + 0.5 \sin(20x). \tag{2.42}$$

In Fig. 2.3 we present the evolution of the system under four different controllers: the greedy boundary control (defined in Proposition 4), the brute force boundary control $u_a = u_b = u^*$, the stabilizing control (defined in Sect. 2.4.3), and the nonlocal control (formula 2.41) with $U = -2$ and T_0 as defined in (2.40). The greedy control allows oscillations to exit from the right boundary but the solution does not converge to the steady state $u^* = -1$. On the other side the brute force control converges to the steady state but generates oscillations on the right boundary as can be seen in Fig. 2.4, top. The stabilizing control also converges but it is less oscillating with respect to the brute force control. The nonlocal control guarantees convergence and avoids oscillations. The evolution of the solution under the action of the stabilizing control and the brute force control are very similar. The decrease of the corresponding Lyapunov function is represented in Fig. 2.4, bottom. One can note how the nonlocal control decreases much faster than the other methods.

To study the dependence of the nonlocal control stabilization performance on the parameters U and T_0 we run several simulations with different values of these parameters, see Fig. 2.5 for $U \in [-2.1, -1.5]$ and $T_0 \in [0.5, 2]$. The convergence time is defined as the first time such that $V(t) \leq 0.1$. We notice that longest convergence time corresponds to $U = -1.5$ and $T_0 = 0.5$ while the fastest corresponds to $U = -2.1$ and $T_0 = 1$. Moreover, for each fixed U there exists an optimal switching time T_0 that minimizes the convergence time.

To further illustrate the oscillations of the boundary trace generated by the brute force control we simulated the case in which the initial datum is strongly oscillating:

$$u_0(x) = 1 + 0.3 \sin(50x). \tag{2.43}$$

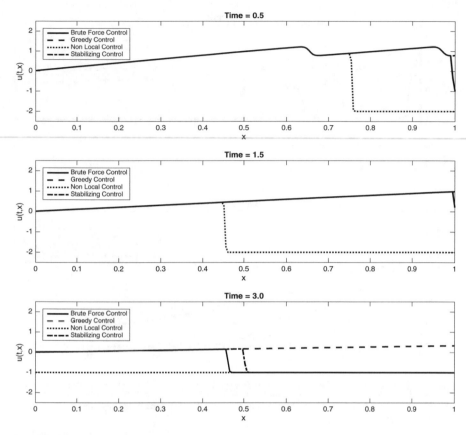

Fig. 2.3 Numerical solution of Burgers equation: Evolution of the solution for the various controllers with oscillating initial data

In Fig. 2.6 the trace of the brute force control shows, at initial times, one big oscillations and then the oscillations continues until $t = 1$. For the stabilizing control the oscillations are smaller but they extend for a longer period up to time $t = 1.5$.

2.5 Mixed Systems PDE-ODE

Here, we present a slightly different system with respect the previous ones of the present Chapter. More precisely, we consider a system of balance laws (PDE) with boundary, coupled with a system of ordinary differential equations (ODE). The coupling condition between the PDE and the ODE is at the level of the boundary. Moreover, we do not consider explicitly a control for such a system. However, the solution to the ODE can be seen as an external control for the system of

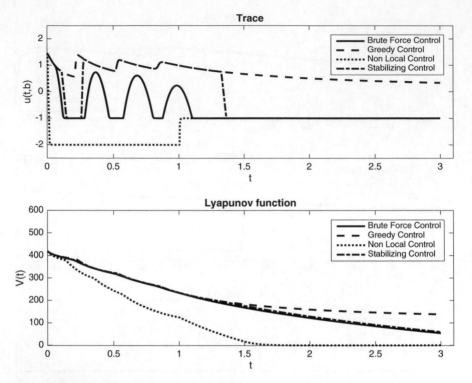

Fig. 2.4 Downstream trace and Lyapunov functions: the Lyapunov functions for different controls are represented in the bottom subfigure. The downstream boundary are represented in the top subfigure respectively

balance laws. The boundary position for the PDE is not fixed a-priori, but it is given by a function γ, which is the solution to an ordinary differential equation. Therefore, we consider the system of conservation laws (2.1) on the domain $\mathcal{D} = \{(t, x) : t \geq 0, \, x \geq \gamma(t)\}$. We present both the mathematical description and a theoretical result about existence of solutions for the Cauchy problem.

Let us consider the following mixed ODE-PDE systems:

$$
\begin{cases}
\partial_t u + \partial_x f(u) = g(u), & x > \gamma(t), t > 0, \\
b\left(u\left(t, \gamma(t)+\right)\right) = B\left(t, w(t)\right), & t > 0, \\
\dot{w} = F\left(t, u\left(t, \gamma(t)+\right), w(t)\right), & t > 0, \\
\dot{\gamma}(t) = \Pi\left(w(t)\right), & t > 0.
\end{cases}
\tag{2.44}
$$

Here the unknowns are $u = u(t, x)$, $w = w(t)$, and $\gamma = \gamma(t)$. As said before, the function u is defined for $t \geq 0$ and $x \geq \gamma(t)$, while w and γ are defined for $t \geq 0$. We assume the following hypotheses.

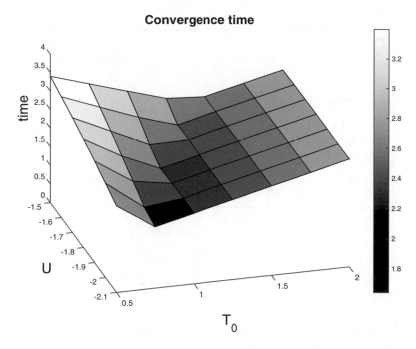

Fig. 2.5 Convergence time of the Lyapunov function: dependence of the convergence time of the Lyapunov function from U and T_0. The convergence time is defined as the first time for which $V(t) \leq 0.1$

(MS.1) $\Omega \subseteq \mathbb{R}^n$ is an open set. Moreover $\hat{u} \in \Omega$, $\hat{w} \in \mathbb{R}^m$ and $\hat{x} \in \mathbb{R}$. For $\delta > 0$, define the sets

$$\mathcal{V} = \left\{ u \in \hat{u} + (\mathbf{BV} \cap \mathbf{L}^1)(\mathbb{R}; \mathbb{R}^n) : u(\mathbb{R}) \subset \Omega \right\},$$

$$\mathcal{V}_\delta = \{ u \in \mathcal{V} : \mathrm{TV}(u) \leq \delta \}.$$

(MS.2) The flux function $f : \Omega \to \mathbb{R}^n$ is smooth. Moreover the system of balance laws is strictly hyperbolic with each characteristic field either genuinely nonlinear or linearly degenerate.

(MS.3) For $\delta > 0$, the source function $g : \mathcal{V}_\delta \to \mathbf{L}^1(\mathbb{R}; \mathbb{R}^n)$ satisfies, for suitable $L_1, L_2 > 0$, the estimates

$$\left\| g(u) - g(u') \right\|_{\mathbf{L}^1} \leq L_1 \left\| u - u' \right\|_{\mathbf{L}^1} \quad \text{and} \quad \mathrm{TV}(g(u)) \leq L_2,$$

for every $u, u' \in \mathcal{V}_\delta$.

(MS.4) $\Pi \in \mathbf{C}^{0,1}(\mathbb{R}^m; \mathbb{R})$.

(MS.5) There exist $c > 0$ and $p \in \{1, 2, \dots, n-1\}$ such that $\lambda_p(\hat{u}) < \Pi(\hat{w}) - c$ and $\lambda_{p+1}(\hat{u}) > \Pi(\hat{w}) + c$.

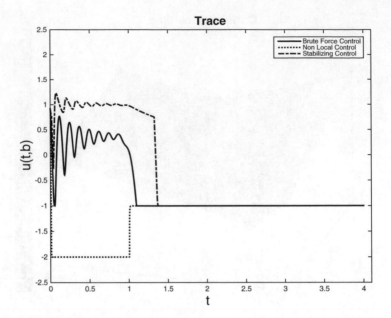

Fig. 2.6 Downstream trace: value of the solution at the right boundary for a strong oscillating initial data

(MS.6) $b \in \mathbf{C}^1(\Omega; \mathbb{R}^{n-p})$ is such that

$$\det \left(D_u b(\hat{u}) \begin{bmatrix} r_{p+1}(\hat{u}) & r_{p+2}(\hat{u}) & \cdots & r_n(\hat{u}) \end{bmatrix} \right) \neq 0.$$

(MS.7)   The map $F : \mathbb{R}^+ \times \Omega \times \mathbb{R}^m \longrightarrow \mathbb{R}^m$ is such that:

   (a) for all $u \in \Omega$ and $w \in \mathbb{R}^m$, the function $t \mapsto F(t, u, w)$ is Lebesgue measurable;
   (b) for all compact subset K of $\Omega \times \mathbb{R}^m$, there exists $C_K > 1$ such that, for all $t \in \mathbb{R}^+$ and $(u_1, w_1), (u_2, w_2) \in K$,

   $$\| F(t, u_1, w_1) - F(t, u_2, w_2) \|_{\mathbb{R}^m} \leq C_K \left(\|u_1 - u_2\|_{\mathbb{R}^n} + \|w_1 - w_2\|_{\mathbb{R}^m} \right) ;$$

   (c) there exists a function $C \in \mathbf{L}^1_{\mathrm{loc}}(\mathbb{R}^+; \mathbb{R}^+)$ such that, for all $t > 0$, $u \in \Omega$ and $w \in \mathbb{R}^m$,

   $$\| F(t, u, w) \|_{\mathbb{R}^m} \leq C(t) \left(1 + \|w\|_{\mathbb{R}^m} \right) .$$

(MS.8) $B \in C^1(\mathbb{R}^+ \times \mathbb{R}^m; \mathbb{R}^{n-p})$ is locally Lipschitz, i.e., for every compact subset K of $\mathbb{R}^m$, there exists a constant $\tilde{C}_K > 0$ such that, for every $t > 0$ and $w \in K$:

$$\left\|\frac{\partial}{\partial t}B(t,w)\right\|_{\mathbb{R}^{n-p}} + \left\|\frac{\partial}{\partial w}B(t,w)\right\|_{\mathbb{R}^{n-p}} \le \tilde{C}_K.$$

Remark 5 Condition **(MS.5)** is analogous to the non-characteristic condition **(NC)** in the case of moving boundary. This is strictly related to conditions **(MS.6)** and **(MS.8)**, describing how boundary conditions are assigned; see also [5–7].

Now we introduce the definition of solution for (2.44).

Definition 4 Let $T > 0$. A triple (u, w, γ) with

$$u \in \mathbf{C}^0\left([0,T]; \mathcal{V}\right) \qquad w \in \mathbf{W}^{1,1}\left([0,T]; \mathbb{R}^m\right) \qquad \gamma \in \mathbf{W}^{1,\infty}\left([0,T]; \mathbb{R}^m\right)$$

is a solution to (2.44) on $[0, T]$ with initial datum (u_0, w_0, x_0) such that $u_0 \in \mathcal{V}$ with $u_0(x) = \hat{u}$ for $x < x_0$, $w_0 \in \mathbb{R}^m$ and $x_0 \in \mathbb{R}$, if

1. u is an entropy admissible solution to

$$\begin{cases} \partial_t u + \partial_x f(u) = g(u), & x > \gamma_*(t), \ t > 0, \\ b\left(u\left(t, \gamma_*(t)+\right)\right) = B_*(t), & t > 0, \end{cases}$$

   on $[0, T]$ with $B_*(t) = B\left(t, w(t)\right)$, $\gamma_*(t) = \gamma(t)$, and initial datum u_0;
2. w solves

$$\begin{cases} \dot{w} = F_*(t, w), & t > 0, \\ w(0) = w_0, \end{cases}$$

   on $[0, T]$ with $F_*(t) = F\left(t, u\left(t, \gamma(t)+\right), w\right)$ a.e.;
3. $\gamma(t) = x_0 + \int_0^t \Pi\left(w(\tau)\right) \, \mathrm{d}\tau$ for a.e. $t \in [0, T]$.

The following result holds; for a proof see [42].

Theorem 19 *Let* **(MS.1)**–**(MS.8)** *hold. Assume that* $b(\hat{u}) = B(0, \hat{w})$. *Then, there exist positive constants* $\delta, \Delta, L, T_\delta$, *domains* $\widehat{\mathcal{D}}_t$ *(for* $t \in [0, T_\delta]$*), and maps*

$$\widehat{P}(t, t_0) \colon \widehat{\mathcal{D}}_{t_0} \to \widehat{\mathcal{D}}_{t_0+t},$$

defined for $t_0, t_0 + t \in [0, T_\delta]$, *such that*

1. $\left(\mathcal{V}_\delta \times B_\delta(\hat{w}) \times \right]\hat{x} - \delta, \hat{x} + \delta[\right) \subseteq \widehat{\mathcal{D}}_t \subseteq \left(\mathcal{V}_\Delta \times B_\Delta(\hat{w}) \times \right]\hat{x} - \Delta, \hat{x} + \Delta[\right)$, *where the notation* $B_r(\hat{w})$ *denotes the ball of radius* r *centered at* $\hat{w}$;
2. *for all* t_0, t_1, t_2 *with* $t_0 \in [0, T_\delta[, t_1 \in [0, T_\delta - t_0[$ *and* $t_2 \in [0, T - t_0 - t_1]$, *we have*

$$\widehat{P}(t_2, t_0 + t_1) \circ \widehat{P}(t_1, t_0) = \widehat{P}(t_1 + t_2, t_0) \qquad \text{and} \qquad \widehat{P}(0, t_0) = Id;$$

3. *for* $t_0 \in [0, T_\delta[, \ t \in [0, T_\delta - t_0], \ and \ (u, w, x), (\bar{u}, \bar{w}, \bar{x}) \in \widehat{\mathcal{D}}_{t_0}$

$$\left\| \widehat{P}(t, t_0)(u, w, x) - \widehat{P}(t, t_0)(\bar{u}, \bar{w}, \bar{x}) \right\|_{\mathbf{L}^1 \times \mathbb{R}^m \times \mathbb{R}}$$

$$\leq L \left(\|u - \bar{u}\|_{\mathbf{L}^1} + \|w - \bar{w}\|_{\mathbb{R}^m} + |x - \bar{x}| \right);$$

4. *for all* $(u_0, w_0, x_0) \in \widehat{\mathcal{D}}_0$, *the map* $t \mapsto \widehat{P}(t, 0)(u_0, w_0, x_0)$, *defined for* $t \in [0, T_\delta]$, *solves* (2.44) *in the sense of Definition 4.*

2.5.1   Examples

We consider here several examples of applications of Theorem 19. In Example 2, we consider the case of a tube with a piston filled with gas, in Example 3 a sewer system with a vertical manhole is considered, in Example 4 we present a system describing a portion of the circulatory system, while in Example 5 the case of a solid body in a fluid is considered.

Example 2 Consider a tube filled with fluid and closed to the left by a piston. The gas dynamics can be described by the *p*-system in the Lagrangian coordinates, coupled with an ordinary differential equation governing the piston's evolution. More precisely

$$\begin{cases} \partial_t \tau - \partial_x v = 0 \\ \partial_t v + \partial_x p(\tau) = 0 \\ V = v(t, 0+) \\ \dot{V} = \alpha \cdot (P(t) - p(\tau(t, 0+))), \end{cases} \tag{2.45}$$

where t is time, x the Lagrangian coordinate (i.e., represent the position of gas particles in the original frame), τ the specific volume, v the Lagrangian speed of the flow, p the pressure in the fluid, V the speed of the piston, $P(t)$ the pressure to the left of the piston, and α is the ratio between the section of the tube and the mass of the piston. The acceleration of the piston is due to the difference between the pressure of the fluid and that of the outer environment. The problem (2.45) can be written in the form (2.44) and, under suitable assumptions, Theorem 19 can be applied.

Example 3 Consider a sewer network composed by a single junction, located at $x = 0$, that joins k horizontal pipes to one vertical manhole. The flow in the i-th tube, for $i = 1, \cdots, n$, can be described by the Saint Venant equations (see [199, formula (108.1)] and Example 8)

$$\begin{cases} \partial_t A_i + \partial_x Q_i = 0 \\ \partial_t Q_i + \partial_x \left(\frac{Q_i^2}{A_i} + p_i(A_i) \right) = 0 \end{cases}$$

ensuring the conservation of mass and momentum. The quantity A_i is the wet cross sectional area, Q_i the flow in the x direction, and p_i is a function representing the hydrostatic pressure. The complete system, which falls within the class (2.44), is

$$\begin{cases} \partial_t A_i + \partial_x Q_i = 0, & i = 1, \cdots, n, \\[2mm] \partial_t Q_i + \partial_x \left(\dfrac{Q_i^2}{A_i} + p_i(A_i) \right) = 0, & i = 1, \cdots, n, \\[3mm] \hat{h}(t) = \dfrac{1}{2g} \dfrac{Q_i(t, 0+)^2}{A_i(t, 0+)^2} + h_i\left(A_i(t, 0+) \right), & i = 1, \cdots, n \\[3mm] h_M(t) = -\dfrac{1}{2g\, A_M^2} \left(\sum_{i=1}^{k} Q_i(t, 0+) \right) \left| \sum_{i=1}^{k} Q_i(t, 0+) \right| + \hat{h}(t), \\[3mm] \dot{h}_M(t) = \dfrac{1}{A_M} \left(Q_{\text{ext}}(t) - \sum_{i=1}^{k} Q_i(t, 0+) \right), \end{cases}$$

$$(2.46)$$

where we require, as boundary condition, the equality of all the hydraulic heads $\hat{h}$ at the junction, and the height h_M of the water inside the manhole is determined by the last two equations of (2.46) based on the conservation of the total amount of water.

Example 4 Following [129, formulæ (2.3), (2.12), (2.14)], [133] and [62], we consider the 1D model for blood flowing through an artery, coupled with a 0D model describing the averaged mass and flow rate in a given terminal compartment of the circulatory system (e.g., capillary bed, venous circulation). The complete model is

$$\begin{cases} \partial_t a + \partial_x q = 0 \\ \partial_t q + \partial_x \left(\alpha \frac{q^2}{a} + \frac{1}{\rho} \pi(a) \right) = -2 \frac{\alpha}{\alpha - 1} \nu \frac{q}{a} \\ a(t, 0+) = a_0 \left(1 + \frac{P - P_{ref}}{\beta} \right)^2 \\ \dot{P} = -\frac{1}{C} Q + \frac{1}{C} q(t, 0) \\ \dot{Q} = -\frac{R}{L} Q + \frac{1}{L} P - \frac{1}{L} P(t, l), \end{cases} \qquad (2.47)$$

where ρ is the blood density, q is the arterial flow rate, a denotes the arterial cross section, $p(a)$ is the arterial blood pressure, $\pi(a) = \int_{a_0}^{a} \tilde{a}\, p'(\tilde{a})\, d\tilde{a}$, and P and Q denote respectively the compartmental mean blood pressure and the compartmental mean flow rate. The remaining constant are:

a_0 reference cross section, ν viscosity coefficient,
α Coriolis coefficient, β arterial wall elasticity,
R compartmental resistance, C compartmental capacitance,
L compartmental inductance.

System (2.47) is of type (2.44).

Example 5 We consider the following model describing the evolution of a solid body, locate at position $\gamma(t)$, inside a compressible fluid, described by the classical p-system (see Example 7):

$$
\begin{cases}
\partial_t \rho + \partial_x q = 0, & x \neq \gamma(t),\ t > 0, \\[2mm]
\partial_t q + \partial_x \left(\dfrac{q^2}{\rho} + p(\rho) \right) = -g\,\rho, & x \neq \gamma(t),\ t > 0, \\[2mm]
\dfrac{q\,(t, \gamma(t)-)}{\rho\,(t, \gamma(t)-)} = \dfrac{q\,(t, \gamma(t)+)}{\rho\,(t, \gamma(t)+)} = V, & t > 0, \\[2mm]
\dot{V} = -g - \dfrac{p\,(\rho\,(t, \gamma(t)+)) - p\,(\rho\,(t, \gamma(t)-))}{m}, & t > 0, \\[2mm]
\dot{\gamma}(t) = V, & t > 0.
\end{cases}
\tag{2.48}
$$

The quantities ρ and q are the fluid mass and linear momentum density above and below the particle, $p = p(\rho)$ is the pressure law, V is the speed of the particle located at $\gamma(t)$ and m is its mass, g is gravity. System (2.48) is a particular case of (2.44).

2.6 Bibliographical Notes

Several papers are concerned with the boundary control of the viscous Burgers equation $\partial_t u + \partial_x \left(\frac{u^2}{2} \right) = \partial_{xx} u$; see [59, 60, 85, 119, 135, 136, 192, 195, 204, 229, 256]. Given a final time $T > 0$, an initial condition u_0 and boundary controls ω_a, ω_b, the control system is

$$
\begin{cases}
\partial_t u + \partial_x \left(\frac{u^2}{2} \right) = \partial_{xx} u \\[2mm]
u(0, x) = u_0(x) \\[2mm]
u(t, a) = \omega_a(t) \\[2mm]
u(t, b) = \omega_b(t).
\end{cases}
\tag{2.49}
$$

When the control acts on a single side (i.e., when ω_a or ω_b is a given preassigned function), there are several results concerning the non-controllability of the system. First, Díaz in [119] proved, using a topological argument, that it is not possible to find a solution u to (2.49), which is arbitrary close to certain open subsets of

$\mathbf{L}^2(a, b)$; see also [135]. In such a case, also the property of exact null controllability does not hold. This means that, for specific initial data u_0, the final profile $u(T, \cdot)$ cannot be constantly equal to 0. In such direction, Fernández-Cara and Guerrero in [130] gave sharp estimates of the minimal time, which depends on the $\mathbf{L}^2$ norm of the initial datum, for which the null controllability property is ensured. Moreover Coron in [86] proved the existence of a time $T > 0$ sufficiently big such that the system (2.49) is null controllable when ω_a (or ω_b) is constantly equal to 0; see also [69].

In case the control acts on both sides, Fursikov and Imanuvilov in [135] proved that every steady state solution can be reached, provided the final time is sufficiently large, whereas Coron in [86] proved that the system can be driven from the null function to every large constant state; see also [151]. In the presence of an additional distributed control, Chapouly in [69] proved the global controllability property for the viscous Burgers' equation; see also [1, 13, 19, 219, 240].

In the case of smooth solutions for general systems of balance laws, problems of exact boundary controllability and of asymptotic stabilization have been addressed. These results were obtained by using explicit formulas for the evolution of the Riemann invariants along characteristics; see [34, 110, 161, 215, 217, 218].

Lyapunov methods for stabilizing classical solutions of 2×2 systems with characteristics speeds of constant opposite sign have been introduced in [90, 122, 241]. Similar work on boundary damping techniques with applications to the Saint-Venant equations has been proposed in [110, 244]. Lyapunov methods for stabilization in the case of solutions with a finite number of shocks are considered in [40]. In the case of hybrid dynamics, see [9].

Mixed systems, composed by hyperbolic conservation laws and ordinary differential equations interacting at the level of the boundary, have been considered in [18, 41–44]. Existence of a solution with the vanishing viscosity approach has been studied in [75].

Chapter 3
Decentralized Control of Conservation Laws on Graphs

3.1 Introduction

Conservation and/or balance laws on networks in the recent years have been the subject of intense study, since a wide range of different applications in real life can be covered by such a research. Among the possible applications, vehicular traffic is probably the one more studied; see, for example, [139] and the reference therein. Other applications range over data networks, irrigation channels, gas pipelines, supply chains, and blood circulation; see, for example, [26, 62, 77, 106, 108, 156]. Therefore, it is natural to consider control problems for such systems, and in particular to consider control functions acting at the level of junctions or nodes [144, 160].

This chapter presents various examples of conservation laws with controls acting at the nodes. More precisely, all the examples are motivated by vehicular traffic and so the dynamics in each edge of the network is described by the Lighthill–Whitham–Richards model [224, 249] or by its discrete version.

In Sect. 3.2 we consider a network composed by a single junction and we assume that the control function modifies the junction Riemann solver; see [144]. A junction Riemann solver is a function which gives a solution to every Riemann problems at the node. In this part, we show that, if the Riemann solver and the control function satisfy a suitable set of theoretical assumptions, then there exists a solution of the corresponding Cauchy problem. The section is completed by three concrete examples of Riemann solvers satisfying the required properties.

In Sect. 3.3, we consider a special case of the problem addressed in the previous ones: the regulation of signalized intersections. More precisely, a traffic light can be seen as a regulation of traffic distribution coefficients, that is of the Riemann solver parameters. The controls in this case are piecewise constant and we address the approximation by continuous ones, providing also precise estimation errors. The approach is strongly based on Moscowitz functions, or cumulative counts of cars flowing at a given point, which solve a Hamilton-Jacobi-Bellman equation.

© The Author(s) 2022
A. Bayen et al., *Control Problems for Conservation Laws with Traffic Applications*,
PNLDE Subseries in Control 99, https://doi.org/10.1007/978-3-030-93015-8_3

In Sect. 3.4, we consider a freeway model with on-ramps and offramps and the controls act at the level of on-ramps mainly through traffic lights; see [248]. In this part, the highway is described by a finite number of consecutive links, and in each link a discretized version of the Lighthill–Whitham–Richards model is considered. Optimal control problems are studied and necessary conditions for optimality are deduced, by computing a generalized gradient of the cost functional.

In Sect. 3.5, we consider a single road with a spatial non-homogeneity located at position $x = x_c$; see [79]. More precisely the aim is the description of toll gates or road works in a limited portion of the road. This is obtained by imposing a variable flux constraint at position $x = x_c$. After giving the definition of solution, we state Theorem 25, which ensure existence and well-posedness of the solution for each admissible control. The section also contains some examples, motivated by applications, of optimal control problems.

Finally, in Sect. 3.6 we consider optimizing the travel time on a loaded network. Solving analytically the problem on a complex network is not feasible thus we resort to optimize some cost function at every junction separately for asymptotic solutions to Riemann Problems. The local problems are solved analytically and the network policies tested numerically. Specifically, we solve analytically the problem for simple junction in Sect. 3.6.1 and simulate the result on two urban networks (in the Italian cities Rome and Salerno) in Sect. 3.6.2. Finally, the case of creating safe corridor for emergency vehicles is addressed in Sect. 3.6.3.

3.2 Control Acting at Nodes Through the Riemann Solver

In this section we consider conservation laws on networks with control acting at the nodes. We refer the reader to Appendix B for notations and basic results of the theory of conservation laws on networks.

We deal with control problems on a node, or more generally on a network with a finite number of arcs and nodes, where the control function ω acts at the level of the node. More precisely, we focus the attention on the following control problem

$$
\begin{cases}
\partial_t u_\ell + \partial_x f(u_\ell) = 0, & \ell \in \{1, \cdots, n+m\}, \\
u_\ell(0, x) = \bar{u}_\ell(x), & \ell \in \{1, \cdots, n+m\}, \, x \in I_\ell, \\
(u_{n+1}(t, 0), \cdots, u_{n+m}(t, 0)) = \mathcal{RS}_{\omega(t)}(u_1(t, 0), \cdots, u_n(t, 0)),
\end{cases}
\tag{3.1}
$$

where $\omega : [0, +\infty[\to \Omega$ is the control function, which acts through the Riemann solver $\mathcal{RS}$.

3.2.1 The Setting of the Problem

Fix a network composed by a finite number of arcs and nodes. Without loss of generality we focus the attention on a single node J with n incoming arcs $I_1, \cdots, I_n$ and m outgoing arcs $I_{n+1}, \cdots, I_{n+m}$; see [142, Theorem 4.3.9] or [139, Section 4.1]. We model each incoming arc I_i $(i \in \{1, \cdots, n\})$ of the node with the real interval $I_i =]-\infty, 0]$. Similarly we model each outgoing arc I_j $(j \in \{n+1, \cdots, n+m\})$ of the node with the real interval $I_j = [0, +\infty[$. On each arc I_ℓ $(\ell \in \{1, \cdots, n+m\})$ we consider the partial differential equation

$$\partial_t u_\ell + \partial_x f(u_\ell) = 0, \tag{3.2}$$

where $u_\ell = u_\ell(t, x) \in [0, u_{max}]$ is the conserved quantity, and f is the flux. For simplicity, we put $u_{max} = 1$.

On the flux f we assume the usual hypothesis

($\mathcal{F}$) $f : [0, 1] \to \mathbb{R}$ is a Lipschitz continuous and concave function satisfying

  (a) $f(0) = f(1) = 0$;
  (b) there exists a unique $\sigma \in]0, 1[$ such that f is strictly increasing in $[0, \sigma[$ and strictly decreasing in $]\sigma, 1]$.

Remark 6 A flux f satisfying condition ($\mathcal{F}$) is not an invertible function. However, if we restrict f on the intervals $[0, \sigma]$ and $[\sigma, 1]$, both restrictions are invertible. It is convenient to define the restrictions

$$f_L : [0, \sigma] \longrightarrow \mathbb{R} \qquad \text{and} \qquad f_R : [\sigma, 1] \longrightarrow \mathbb{R} \tag{3.3}$$
$$\quad u \longmapsto f(u) \qquad\qquad\qquad u \longmapsto f(u).$$

As said before both f_L and f_R are invertible.

The concept of solution to (3.2) is the following.

Definition 5 Fix $\ell \in \{1, \cdots, n+m\}$. A function $u_\ell \in \mathbf{C}^0([0, +\infty[; \mathbf{L}^1_{loc}(I_\ell))$ is an entropy admissible solution to (3.2) in the arc I_ℓ if, for every $k \in [0, 1]$ and every $\tilde{\varphi} : [0, +\infty[\times I_\ell \to \mathbb{R}$ smooth, positive with compact support contained in $]0, +\infty[\times (I_\ell \setminus \{0\})$, it holds:

$$\int_0^{+\infty} \int_{I_\ell} \left[|u_\ell - k| \, \partial_t \tilde{\varphi} + \mathrm{sgn}(u_\ell - k) \, (f(u_\ell) - f(k)) \, \partial_x \tilde{\varphi} \right] dx \, dt \geq 0. \tag{3.4}$$

Let us introduce the concept of admissible control functions.

Definition 6 We say that $\omega : [0, +\infty[\to \Omega$ is an admissible control function at the node J if:

1. Ω is a subset of a normed space $(\overline{\Omega}, \|\cdot\|_{\overline{\Omega}})$;
2. Ω is the finite union of connected and pairwise disjoint sets;

3. ω is a right continuous function with finite total variation;
4. there exists a finite number of jumps between the connected components of Ω.

The definition of Riemann solver at a node is the following one.

Definition 7 A Riemann solver $\mathcal{RS}$ at J is a function

$$\mathcal{RS}: \quad [0,1]^{n+m} \quad \longrightarrow \quad [0,1]^{n+m}$$
$$(u_{1,0}, \cdots, u_{n+m,0}) \longmapsto (\bar{u}_1, \cdots, \bar{u}_{n+m})$$

satisfying the following properties:

1. $\sum_{i=1}^{n} f_i(\bar{u}_i) = \sum_{j=n+1}^{n+m} f_j(\bar{u}_j)$;
2. for every $i \in \{1, \cdots, n\}$, the classical Riemann problem

$$\begin{cases} \partial_t u + \partial_x f(u) = 0, & x \in \mathbb{R}, \, t > 0, \\ u(0, x) = \begin{cases} u_{i,0}, & \text{if } x < 0, \\ \bar{u}_i, & \text{if } x > 0, \end{cases} \end{cases}$$

   is solved with waves with negative speed;
3. for every $j \in \{n+1, \cdots, n+m\}$, the classical Riemann problem

$$\begin{cases} \partial_t u + \partial_x f(u) = 0, & x \in \mathbb{R}, \, t > 0, \\ u(0, x) = \begin{cases} \bar{u}_j, & \text{if } x < 0, \\ u_{j,0}, & \text{if } x > 0, \end{cases} \end{cases}$$

   is solved with waves with positive speed;
4. the consistency condition

$$\mathcal{RS}(\mathcal{RS}(u_{1,0}, \cdots, u_{n+m,0})) = \mathcal{RS}(u_{1,0}, \cdots, u_{n+m,0})$$

   holds.

Fix a family $(\mathcal{RS}_p)_{p \in \Omega}$ of Riemann solvers at the node J. The concept of solution to (3.1) is given by the following definition.

Definition 8 A collection of functions $u_\ell \in \mathbf{C}^0([0, +\infty[; \mathbf{L}^1_{loc}(I_\ell))$, provides a solution at J to (3.1) if

1. for every $\ell \in \{1, \cdots, n+m\}$, the function u_ℓ is an entropy admissible solution to (3.2) in the arc I_ℓ;
2. for every $\ell \in \{1, \cdots, n+m\}$ and for a.e. $t > 0$, the function $x \mapsto u_\ell(t, x)$ has a version with bounded total variation;
3. for a.e. $t > 0$, it holds

$$\mathcal{RS}_{\omega(t)} (u_1(t, 0), \cdots, u_{n+m}(t, 0)) = (u_1(t, 0), \cdots, u_{n+m}(t, 0)), \qquad (3.5)$$

where u_ℓ stands for the version with bounded total variation of 2.

3.2.2   The Main Result

Here we state the main result, which deals with the existence of a solution to (3.1). As a preliminary, we need to introduce properties (P1)-(P4) for a family of Riemann solvers $(\mathcal{RS}_p)_{p\in\Omega}$. These properties ensure some bounds on approximate wave-front tracking solutions, used to prove Theorem 20.

Definition 9 We say that the family of Riemann solvers $(\mathcal{RS}_p)_{p\in\Omega}$ has the property (P1) if the following condition holds. Given $(u_{1,0}, \cdots, u_{n+m,0})$ and $(u'_{1,0}, \cdots, u'_{n+m,0})$ two initial data such that $u_{\ell,0} = u'_{\ell,0}$ whenever either $u_{\ell,0}$ or $u'_{\ell,0}$ is a bad datum, then

$$\mathcal{RS}_p(u_{1,0}, \cdots, u_{n+m,0}) = \mathcal{RS}_p(u'_{1,0}, \cdots, u'_{n+m,0}) \tag{3.6}$$

for every $p \in \Omega$.

Property (P2) asks for bounds in the increase of the flux variation for waves interacting with J. More precisely the latter should be bounded in terms of the strength of the interacting wave as well as the variation in the incoming fluxes.

Definition 10 We say that the family of Riemann solvers $(\mathcal{RS}_p)_{p\in\Omega}$ has the property (P2) if there exists a constant $C \geq 1$ such that the following condition holds. For every $p \in \Omega$, for every equilibrium $(u_{1,0}, \cdots, u_{n+m,0})$ of $\mathcal{RS}_p$ and for every wave $(u_{\ell,0}, u_\ell)$ $(\ell \in \{1, \cdots, n+m\})$ interacting with J at time $\bar{t} > 0$ and producing waves in the arcs according to the Riemann solver $\mathcal{RS}_p$, we have

$$\begin{aligned} \mathrm{TV}_f(\bar{t}+) - \mathrm{TV}_f(\bar{t}-) \\ \leq C \min\{|f(u_{\ell,0}) - f(u_\ell)|, |\Gamma(\bar{t}+) - \Gamma(\bar{t}-)|\}. \end{aligned} \tag{3.7}$$

The property (P3) states that a wave interacting with J with a flux decrease on a specific arc should also give rise to a decrease in the incoming fluxes.

Definition 11 We say that the family of Riemann solvers $(\mathcal{RS}_p)_{p\in\Omega}$ has the property (P3) if, for every $p \in \Omega$, for every equilibrium $(u_{1,0}, \cdots, u_{n+m,0})$ of $\mathcal{RS}_p$ and for every wave $(u_{\ell,0}, u_\ell)$ $(\ell \in \{1, \cdots, n+m\})$ with $f(u_\ell) < f(u_{\ell,0})$, interacting with J at time $\bar{t} > 0$ and producing waves in the arcs according to the Riemann solver $\mathcal{RS}_p$, we have

$$\Gamma(\bar{t}+) \leq \Gamma(\bar{t}-). \tag{3.8}$$

Finally property (P4) describes the variation of the fluxes due to a variation of the parameter of the Riemann solver.

Definition 12 We say that the family of Riemann solvers $(\mathcal{RS}_p)_{p\in\Omega}$ has the property (P4) if there exists $C > 0$ such that, for every p_1, p_2 in the same connected component of Ω and for every equilibrium $(u_{1,0}, \cdots, u_{n+m,0})$ for $\mathcal{RS}_{p_1}$ we have

$$\sum_{\ell=1}^{n+m} \left| f(\hat{u}_\ell) - f(u_{\ell,0}) \right| \leq C \, \| p_1 - p_2 \|_P \,, \tag{3.9}$$

where $(\hat{u}_1, \cdots, \hat{u}_{n+m}) = \mathcal{RS}_{p_2}(u_{1,0}, \cdots, u_{n+m,0})$.

The following result holds.

Theorem 20 *Suppose that ω is an admissible control function in the sense of Definition 6. Assume that, for every $\ell \in \{1, \cdots, n + m\}$, $u_\ell \in \mathbf{L}^1 (I_\ell; [0, 1])$ with finite total variation. Assume moreover that the family of Riemann solvers $(\mathcal{RS}_p)_{p \in \Omega}$ satisfies properties (P1)–(P4).*

Then there exists a solution $(u_1, \cdots, u_{n+m})$ to the Cauchy problem (3.1) in the sense of Definition 8.

The proof is based on the wave-front tracking technique; see [144] or [139].

3.2.3 Example of Family of Riemann Solvers

Aim of this part is to present different examples of family of Riemann solvers satisfying properties (P1), (P2), (P3), and (P4).

3.2.3.1 The Riemann Solver $\mathcal{RS}_1$

This example is based on the Riemann solver introduced for vehicular traffic in [76]. First introduce the set of matrices

$$\mathcal{A} := \left\{ A = \{a_{ji}\} \begin{array}{c} i=1,\cdots,n \\ j=n+1,\cdots,n+m \end{array} : \begin{array}{c} 0 < a_{ji} < 1 \; \forall i, j, \\ \displaystyle\sum_{j=n+1}^{n+m} a_{ji} = 1 \; \forall i \end{array} \right\}. \tag{3.10}$$

Let $\{e_1, \cdots, e_n\}$ be the canonical basis of $\mathbb{R}^n$. For every $i = 1, \cdots, n$, we denote $H_i = \{e_i\}^\perp$. If $A \in \mathcal{A}$, then we write, for every $j = n + 1, \cdots, n + m$, $a_j = (a_{j1}, \cdots, a_{jn}) \in \mathbb{R}^n$ and $H_j = \{a_j\}^\perp$. Introduce now the following notation for sets of indices:

- let $\mathcal{H}$ be the set of indices $\varsigma = (\varsigma_1, \cdots, \varsigma_n)$ such that $\varsigma_i \in \mathbb{N}$ for every $i \in \{1, \cdots, n\}$ and $1 \leq \varsigma_1 < \cdots < \varsigma_n \leq n + m$;
- let $\mathcal{K}$ be the set of indices $\mathbf{k} = (k_1, \ldots, k_\ell)$ such that $\ell \in \{1, \cdots, n - 1\}$, $k_i \in \mathbb{N}$ for every $i \in \{1, \cdots, \ell\}$ and $1 \leq k_1 < k_2 < \cdots < k_\ell \leq n + m$.

Writing $\mathbf{1} = (1, \cdots, 1) \in \mathbb{R}^n$, for every $\mathbf{k} \in \mathcal{K}$ define

$$H_{\mathbf{k}} = \bigcap_{h=1}^{\ell} H_{k_h}$$

and the set

$$\mathfrak{N} := \left\{ A \in \mathcal{A} : \mathbf{1} \notin H_{\mathbf{k}}^{\perp} \text{ for every } \mathbf{k} \in \mathcal{K} \right\}. \tag{3.11}$$

Moreover, given $0 < \kappa_1 < \kappa_2 < 1$, define

$$\mathfrak{N}_{\kappa_1}^{\kappa_2} = \left\{ A \in \mathfrak{N} : \kappa_1 \le a_{ji} \le \kappa_2, \forall i = 1, \cdots, n, \forall j = n+1, \cdots, n+m \right\}. \tag{3.12}$$

The construction of the Riemann solver $\mathcal{RS}1$ can be summarized as follows.

1. Fix a matrix $A \in \mathfrak{N}$ and consider the closed, convex, and not empty set

$$\Omega = \left\{ (\gamma_1, \cdots, \gamma_n) \in \prod_{i=1}^{n} \Omega_i : A \cdot (\gamma_1, \cdots, \gamma_n)^T \in \prod_{j=n+1}^{n+m} \Omega_j \right\}. \tag{3.13}$$

2. Find the point $(\bar{\gamma}_1, \cdots, \bar{\gamma}_n) \in \Omega$ which maximizes the function

$$E(\gamma_1, \cdots, \gamma_n) = \gamma_1 + \cdots + \gamma_n, \tag{3.14}$$

and define $(\bar{\gamma}_{n+1}, \cdots, \bar{\gamma}_{n+m})^T := A \cdot (\bar{\gamma}_1, \cdots, \bar{\gamma}_n)^T$. Since $A \in \mathfrak{N}$, the point $(\bar{\gamma}_1, \cdots, \bar{\gamma}_n)$ is uniquely defined.

3. For every $i \in \{1, \cdots, n\}$, set $\bar{u}_i$ either by $u_{i,0}$ if $f(u_{i,0}) = \bar{\gamma}_i$, or by the solution to $f(u) = \bar{\gamma}_i$ such that $\bar{u}_i \ge \sigma_i$. For every $j \in \{n+1, \cdots, n+m\}$, set $\bar{u}_j$ either by $u_{j,0}$ if $f(u_{j,0}) = \bar{\gamma}_j$, or by the solution to $f(u) = \bar{\gamma}_j$ such that $\bar{u}_j \le \sigma_j$. Finally, define $\mathcal{RS}_{1,A} : [0,1]^{n+m} \to [0,1]^{n+m}$ by

$$\mathcal{RS}_{1,A}(u_{1,0}, \cdots, u_{n+m,0}) = (\bar{u}_1, \cdots, \bar{u}_n, \bar{u}_{n+1}, \cdots, \bar{u}_{n+m}). \tag{3.15}$$

In this way we have defined a family of Riemann solvers $\mathcal{RS}_{1,A}$ depending on the matrix $A \in \mathfrak{N}$. For a proof that such a family of Riemann solvers satisfies properties (P1), (P2), (P3), and (P4) see [144] or [139, Chapter 4].

3.2.3.2 The Riemann Solver $\mathcal{RS}_2$

This example is based on the Riemann solver introduced for telecommunication networks in [108]; see also [139, Chapter 4.2.2]. Consider the set

$$\Theta = \left\{ \boldsymbol{\theta} = (\theta_1, \cdots, \theta_{n+m}) \in \mathbb{R}^{n+m} : \begin{array}{l} \theta_1 > 0, \cdots, \theta_{n+m} > 0, \\ \sum_{i=1}^{n} \theta_i = \sum_{j=n+1}^{n+m} \theta_j = 1 \end{array} \right\} \tag{3.16}$$

of vectors $\boldsymbol{\theta}$, whose components are right of way parameters and so describe the relative importance of the edges of the node. Note that Θ is a convex subset of $\mathbb{R}^{n+m}$; also it is arc-wise connected and so it is a connected subset of $\mathbb{R}^{n+m}$.

The Riemann solver $\mathcal{RS}_2$ can be constructed with the following steps.

1. Fix $\boldsymbol{\theta} \in \Theta$ and define

$$\Gamma_{inc} = \sum_{i=1}^{n} \sup \Omega_i, \quad \Gamma_{out} = \sum_{j=n+1}^{n+m} \sup \Omega_j,$$

then the maximal possible through-flow at the crossing is

$$\Gamma = \min \{\Gamma_{inc}, \Gamma_{out}\} .$$

2. Introduce the closed, convex, and not empty sets

$$I = \left\{ (\gamma_1, \cdots, \gamma_n) \in \prod_{i=1}^{n} \Omega_i : \sum_{i=1}^{n} \gamma_i = \Gamma \right\}$$

$$J = \left\{ (\gamma_{n+1}, \cdots, \gamma_{n+m}) \in \prod_{j=n+1}^{n+m} \Omega_j : \sum_{j=n+1}^{n+m} \gamma_j = \Gamma \right\} .$$

3. Denote with $(\bar{\gamma}_1, \cdots, \bar{\gamma}_n)$ the orthogonal projection on the convex set I of the point $(\Gamma\theta_1, \cdots, \Gamma\theta_n)$ and with $(\bar{\gamma}_{n+1}, \cdots, \bar{\gamma}_{n+m})$ the orthogonal projection on the convex set J of the point $(\Gamma\theta_{n+1}, \cdots, \Gamma\theta_{n+m})$.
4. For every $i \in \{1, \cdots, n\}$, define $\bar{u}_i$ either by $u_{i,0}$ if $f(u_{i,0}) = \bar{\gamma}_i$, or by the solution to $f(u) = \bar{\gamma}_i$ such that $\bar{u}_i \geq \sigma_i$. For every $j \in \{n+1, \cdots, n+m\}$, define $\bar{u}_j$ either by $u_{j,0}$ if $f(u_{j,0}) = \bar{\gamma}_j$, or by the solution to $f(u) = \bar{\gamma}_j$ such that $\bar{u}_j \leq \sigma_j$. Finally, define $\mathcal{RS}_{2,\theta} : [0, 1]^{n+m} \to [0, 1]^{n+m}$ by

$$\mathcal{RS}_{2,\theta}(u_{1,0}, \cdots, u_{n+m,0}) = (\bar{u}_1, \cdots, \bar{u}_n, \bar{u}_{n+1}, \cdots, \bar{u}_{n+m}) . \qquad (3.17)$$

For a proof that such a family of Riemann solvers satisfies properties (P1), (P2), (P3), and (P4) see [144] or [139, Chapter 4].

3.2.3.3 The Riemann Solver $\mathcal{RS}_3$

This example is based on the Riemann solver introduced for car traffic in [232] for modeling T-nodes. Consider a node J with n incoming and $m = n$ outgoing arcs and fix a positive coefficient Γ_J, which represents the maximum capacity of the node. The construction of the Riemann solver can be done in the following way.

1. Fix $\boldsymbol{\theta} \in \Theta$, where Θ is defined in (3.16). For every $i \in \{1, \cdots, n\}$, define

$$\Gamma_i = \min\left\{\gamma_i^{max}, \gamma_{i+n}^{max}\right\},$$

then the maximal possible through-flow at J is

$$\Gamma = \sum_{i=1}^{n} \Gamma_i.$$

2. Introduce the closed, convex, and not empty set

$$I = \left\{(\gamma_1, \cdots, \gamma_n) \in \prod_{i=1}^{n}[0, \Gamma_i] : \sum_{i=1}^{n} \gamma_i = \min\{\Gamma, \Gamma_J\}\right\}.$$

3. Denote with $(\bar{\gamma}_1, \cdots, \bar{\gamma}_n)$ the orthogonal projection on the convex set I of the point $(\min\{\Gamma, \Gamma_J\}\theta_1, \cdots, \min\{\Gamma, \Gamma_J\}\theta_n)$ and set $(\bar{\gamma}_{n+1}, \cdots, \bar{\gamma}_{2n}) = (\bar{\gamma}_1, \cdots, \bar{\gamma}_n)$.
4. For every $i \in \{1, \cdots, n\}$, define $\bar{u}_i$ either by $u_{i,0}$ if $f(u_{i,0}) = \bar{\gamma}_i$, or by the solution to $f(u) = \bar{\gamma}_i$ such that $\bar{u}_i \geq \sigma_i$. For every $j \in \{n + 1, \cdots, n + m\}$, define $\bar{u}_j$ either by $u_{j,0}$ if $f(u_{j,0}) = \bar{\gamma}_j$, or by the solution to $f(u) = \bar{\gamma}_j$ such that $\bar{u}_j \leq \sigma_j$. Finally, define $\mathcal{RS}_{3,\boldsymbol{\theta}} : [0, 1]^{n+m} \to [0, 1]^{n+m}$ by

$$\mathcal{RS}_{3,\boldsymbol{\theta}}(u_{1,0}, \cdots, u_{n+m,0}) = (\bar{u}_1, \cdots, \bar{u}_n, \bar{u}_{n+1}, \cdots, \bar{u}_{n+m}). \qquad (3.18)$$

In this way we have defined a family of Riemann solvers $\mathcal{RS}_{3,\boldsymbol{\theta}}$ depending on the parameter $\boldsymbol{\theta}$. For a proof that such a family of Riemann solvers satisfies properties (P1), (P2), (P3), and (P4) see [144] or [139, Chapter 4].

3.3 Modeling Signalized Intersections

This section is devoted to a specific problem of choosing traffic distribution coefficients: the regulation of signalized intersections. Some technical results will be stated without proofs, referring the reader to [165, 166].

Notice that a traffic signal can be interpreted as a special case of (3.1) with the control signal ω taking values in a discrete set, e.g., $\{green, red\}$, and being piecewise constant. The main interest here is to approximate the problem with a continuous one, where the controls represent the traffic distribution coefficients corresponding to a given signal schedule.

The main idea for continuous approximation is as follows. Fix a simple junction with two incoming roads I_1 and I_2 and one outgoing I_3. The main parameter of the problem is the fraction, say $\eta \in (0, 1)$, of the signal cycle for which I_1 has a green signal, so that $1 - \eta$ is the fraction of red signal for I_1. Then the continuous approximation corresponds to assign a priority to I_1 so that the resulting fraction of the whole traffic flowing to I_3, which is coming from I_1, is equal to η. For a fixed time horizon $[0, T]$, all signal controls will be represented by periodic functions $\omega_i(\cdot) : [0, T] \to \{0, 1\}$ such that $\omega_i(t)$ equals one if the signal is green and equals zero otherwise. We denote by $\Delta_i \in \mathbb{R}_+$ the cycle length, i.e., the period of u_i, assume it starts with green and let η_i denote the green fraction.

We parametrize road I_i with the interval $[a_i, b_i] \subset \mathbb{R}$, denote by $f_i(u)$ the fundamental diagram (i.e., flux function), by u_i^c the critical density, where the flow is maximal $C_i = f_i(u_i^c)$, and by u^{jam} the maximal or jam density. The density on road I_i is denoted by $u_i(t, x)$, for $(t, x) \in [0, T] \times [a_i, b_i]$. The *demand* $D_i(t)$ and *supply* $S_i(t)$ of the road are given by:

$$D_i(t) = \begin{cases} C_i & \text{if } u_i(t, b_i-) \geq u_i^c \\ f_i\big(u_i(t, b_i-)\big) & \text{if } u_i(t, b_i-) < u_i^c \end{cases} \tag{3.19}$$

$$S_i(t) = \begin{cases} C_i & \text{if } u_i(t, a_i+) < u_i^c \\ f_i\big(u_i(t, a_i+)\big) & \text{if } u_i(t, a_i+) \geq u_i^c; \end{cases} \tag{3.20}$$

The notion of *effective supply* is of crucial importance to the articulation of signal models.

Definition 13 (Effective Supply) Given any link I_i, with downstream links $\{I_j : j = 1, 2, \cdots, m_i\}$, the effective supply for I_i is defined as

$$\mathcal{E}_i(t) \doteq \min\left\{ C_i, \min_{j=1,\cdots,m_i}\left\{ \frac{S_j(t)}{\alpha_{i,j}(t)} \right\} \right\}, \tag{3.21}$$

where $\alpha_{i,j}(t) \in \mathbb{R}^+$, satisfying $\sum_{j=1}^{m_i} \alpha_{i,j}(t) \equiv 1$, are the car turning percentages, C_i is the flow capacity of the link I_i, and S_j is the supply function for I_j.

The time-varying quantity $\mathcal{E}_i(t)$ expresses the downstream receiving capacity available for I_i if no signal controls is present. The superscripts "Δ" and "0" represent quantities associated with the on-and-off signal model and the continuum signal model, respectively. For given signal control ω_i and effective supply $\mathcal{E}_i(t)$, the on-and-off model is expressed in terms of its downstream boundary condition:

$$f_{out,i}^{\Delta}(t) = f_i(u_i(t, b_i)) = \min\left\{ D_i(t), \omega_i(t)\mathcal{E}_i(t) \right\} \quad \text{(On-and-off)}, \tag{3.22}$$

where $f_{out,i}^{\Delta}(t)$ is the exit flow, while the continuum signal model is given by:

$$f_{out,i}^{0}(t) = f_i(u_i(t, b_i)) = \min\left\{ D_i(t), \eta_i\mathcal{E}_i(t) \right\} \quad \text{(Continuum)}. \tag{3.23}$$

3.3.1 The Hamilton-Jacobi Representation of Signal Models

Denote by $N(t, x)$ the Moskowitz function, i.e., the number of vehicles passed by location x before time t. The function $N(t, x)$ satisfies the following Hamilton-Jacobi equation

$$\partial_t N(t, x) - f\left(-\partial_x N(t, x)\right) = 0 \qquad (t, x) \in [0, T] \times [a, b] \qquad (3.24)$$

subject to initial condition and boundary conditions. We use a semi-analytic solution representation of the Hamilton-Jacobi equation (3.24), namely the generalized Lax-Hopf formula (see [21, 72]). To isolate a unique solution, we specify the initial condition $N_{ini}(x)$, the upstream boundary condition $N_{up}(t)$, and the downstream boundary condition $N_{down}(t)$ and the weak downstream boundary condition

$$V^\Delta(t) \doteq \int_0^t u(\tau) \cdot \mathcal{E}(\tau) \, d\tau, \qquad V^0(t) \doteq \int_0^t \eta \mathcal{E}(\tau) \, d\tau. \qquad (3.25)$$

For the same initial and upstream boundary conditions, the difference in the solution $N(t, x)$ is bounded by the difference in the weak downstream boundary conditions:

$$\max_{t \in [0, \tau], x \in [a, b]} \left| N^\Delta(t, x) - N^0(t, x) \right| \leq \max_{t \in [0, t]} \left| V^\Delta(t) - V^0(t) \right|, \qquad (3.26)$$

where $N^\Delta(t, x)$ and $N^0(t, x)$ represent the Moskowitz function corresponding to the on-and-off and the continuum models, respectively. Our aim is to compare the two signal models studying the convergence of the on-and-off model to the continuum model and determining an error bound on the continuum approximation error. For simplicity let us focus on the simple merge network depicted in Fig. 3.1, with incoming roads I_1, I_2 and outgoing ones I_3. Let $\omega_1(t) \in \{0, 1\}$ and $\omega_2(t) \in \{0, 1\}$ be the on-and-off signals at the merge junction A, while ω_3 the one at location B. The presence of the second signal allows to study the effect of the spillback presented in Sect. 3.3.3. We focus on road I_1 being the analysis for I_2 similar.

Fig. 3.1 A signalized merge junction

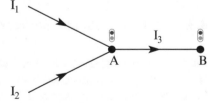

3.3.2 When Spillback Is Absent

We first focus on the case of no spillback at the merge junction A. Our analysis is valid for all fundamental diagrams which satisfy the following condition:

(F)   the fundamental diagram $f(u)$ of each link is continuous and concave, and vanishes at $u = 0$ and $u = u^{jam}$.

The lack of spillback at A implies that road I_3 remains in free flow state, therefore the supply function $S_3(t)$ of I_3 is equal to its flow capacity C_3. The convergence result is given by the following:

Theorem 21 *Consider the merge junction of Fig. 3.1, a signal control $\omega_1(\cdot)$ (for link I_1) with cycle length (period of ω_1) indicated by Δ_A, and split parameter (i.e., green-red ratio) $\eta_1 \in (0, 1)$. Let $N_{up}(t)$ and $N_{ini}(x)$ be the upstream boundary condition and initial condition for (3.24) and $N^{\Delta_A}(t, x)$ and $N^0(t, x)$ be the solutions with additional weak downstream boundary conditions $V^{\Delta_A}(t)$ and $V^0(t)$ respectively, given in (3.25). If the entrance of link I_3 remains in the uncongested phase, then $N^{\Delta_A}(t, x) \to N^0(t, x)$ uniformly for all $(t, x) \in [0, T] \times [a_1, b_1]$, as $\Delta_A \to 0$.*

For practical purpose it is important to provide explicit formulas to estimate the error between the solutions produced by continuous and on-and-off signals. We have the following:

Theorem 22 (Error Estimate Without Spillback) *Consider the merge junction of Fig. 3.1 and conditions $N_{ini}(x)$ and $N_{up}(t)$ for (3.24) on road I_1. Let $N^{\Delta_A}(t, x)$ and $N^0(t, x)$ be the solutions with additional downstream boundary conditions $V^{\Delta_A}(t)$ and $V^0(t)$, given in (3.25). If the entrance of link I_3 remains in the uncongested phase, then for all $(t, x) \in [0, T] \times [a_1, b_1]$,*

$$\left| N^{\Delta_A}(t, x) - N^0(t, x) \right| \leq \eta_1 (1 - \eta_1) \Delta_A \min\{C_1, C_3\}$$

$$\leq \frac{1}{4} \Delta_A \min\{C_1, C_3\}. \tag{3.27}$$

3.3.3 When Spillback Is Present and Sustained

We turn now to consider the situation when spillback occurs at merge junction A and is sustained (i.e., not absorbed back by the junction for a large number of cycles). Such situation is called *sustained spillback* and is very common when the demand of roads I_1 and I_2 is at high levels and not me by the supply of I_3.

First we show analytically the lack of convergence for the case of a triangular fundamental diagram:

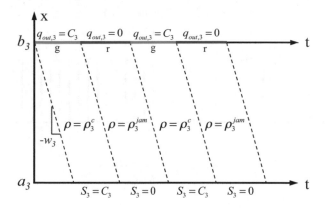

Fig. 3.2 Wave dynamics on I_3 in congested phase for a triangular fundamental diagram. The dashed lines represent characteristics traveling backward at speed w_3

$$f(u) = \begin{cases} vu & \text{if } u \in [0, u^c] \\ -w(u - u^{jam}) & \text{if } u \in (u^c, u^{jam}], \end{cases} \qquad (3.28)$$

where v, respectively w, is the speed of the forward, respectively backward, propagating waves. Notice that V is also the free flow speed of cars.

Let us first examine the dynamics on road I_3. Since I_3 is in congested phase, the characteristic lines with slope $-w_3$ (the backward wave speed on I_3) emit from the right boundary $x = b_3$ and reach the left boundary $x = a_3$ (see Fig. 3.2). When the light is red, the flow q_3 exiting I_3 vanishes, producing a jam density value u_3^{jam}, while when the light is green q_3 is equal to the flow capacity C_3 generating a density value u_3^c (the critical density on I_3). As a result, the supply function $S_3(t)$ at the entrance of I_3 fluctuates between 0 and C_3, leading $\mathcal{E}_1(t)$ to fluctuate between 0 and $\min\{C_1, C_3\}$.

The effective supply $\mathcal{E}_1(t)$ does not have bounded variation as we send the signal cycle of $\omega_3(t)$ to zero. An example is obtained using resonant signals, i.e., ω_1 and ω_3 such that $S_3(t) = C_3 \cdot \omega_1(t)$. We get:

$$\int_0^t \mathcal{E}_1(\tau)\omega_1(\tau)\, d\tau = \int_0^t \min\{C_1, C_3\} \cdot \omega_1^2(\tau)\, d\tau$$

$$= \int_0^t \min\{C_1, C_3\} \cdot \omega_1(\tau)\, d\tau = \int_0^t \mathcal{E}_1(\tau)\, d\tau$$

which does not converge to $\eta_1 \int_0^t \mathcal{E}_1(\tau)\, d\tau$ (regardless of the cycle length Δ_A).

The case of strictly concave fundamental diagram is completely different and we can establish convergence. Assume that f is a piecewise smooth function that satisfies, in addition to **(F)**,

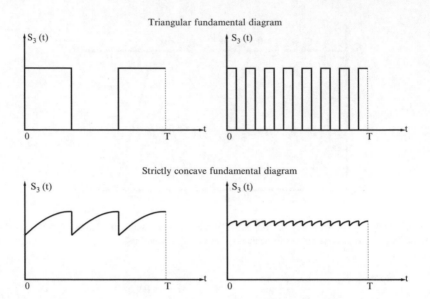

Fig. 3.3 Supply observed at the entrance of I_3, First row: the triangular case; second row: the strictly concave case. First column: larger signal cycle; second column: smaller signal cycle

$$f''(u) \leq -b \qquad \text{for some } b > 0 \qquad\qquad (3.29)$$

for all $u \in [0, u^{jam}]$ such that f is twice differentiable at u. Because of the genuine nonlinearity of the characteristic field, any waves bearing a flux variation, generated by signal control at the exit of the link, is reduced while propagating backward. We are not ready to state our main result, and refer the reader to the papers [165, 166] for detailed proof:

Lemma 2 *Consider the merge junction of Fig. 3.1, assume I_3 parameterized by $[0, L]$ and a strictly concave fundamental diagram f. Then the supply function $S_3(t)$ converges to some constant S_3^* uniformly as the cycle length of signal $\omega_3(t)$ tends to zero (Fig. 3.3).*

The main result for strictly concave case is as follows:

Theorem 23 *Consider a network with a fixed-cycle-and-split signal control at each node with a strictly concave fundamental diagram. Then the solution of this network converges to the one corresponding to the continuum signal model, when the traffic signal cycles tend to zero.*

For the error estimation we have the following:

Theorem 24 (Error Estimate with Sustained Spillback) *Consider the setting and notations of Theorem 22, with road I_3 subject to sustained spillback. If the fundamental diagram of I_3 is triangular, then*

$$\left| N^{\Delta_A}(t, x) - N^0(t, x) \right| \leq \eta_1 (1 - \eta_1) \Delta_A \min\{C_1, C_3\}$$
$$+ \min\{C_1, C_3\} \eta_1 t, \tag{3.30}$$

where C_1 and C_3 denotes the flow capacity of link I_1 and I_3 respectively. If the fundamental diagram of I_3 is strictly concave, then

$$\left| N^{\Delta_A}(t, x) - N^0(t, x) \right| \leq \min \left\{ C_1, f\left((f')^{-1} \left(\frac{-L}{L/w + \Delta_B} \right) \right) \right\} \eta_1 t$$
$$+ \eta_1 (1 - \eta_1) \Delta_A \min\{C_1, C_3\} \tag{3.31}$$

for all $(t, x) \in [0, T] \times [a_1, b_1]$, where $N^{\Delta_A}(t, x)$ and $N^0(t, x)$ are the Moskowitz functions with the on-and-off model and the continuum model, respectively. Here Δ_B denotes the cycle length of $\omega_3(t)$.

The next proposition is an immediate consequence of (3.31) and provides useful information on the accuracy of the continuous approximation:

Proposition 5 *Assume a strictly concave fundamental diagram f for I_3. When spillback occurs, the approximation error $|N^{\Delta}(t, x) - N^0(t, x)|$ decreases with larger length L_3 of I_3, and/or with smaller signal cycle Δ_B. Moreover, the size of the error is determined only by the congested branch of the fundamental diagram.*

3.4 Control for a Freeway Model

This section deals with control problems in case of a freeway model with on-ramps and offramps. This kind of problem has been studied by several authors; see, for example, [153, 247, 248] and the references therein.

More precisely, we fix a terminal time T and we consider a finite-horizon control problem on the time interval $[0, T]$.

3.4.1 Freeway Model

We consider a freeway road and we model it with a sequence of N piece of roads, called links and labeled by an index $\ell \in \{1, \ldots, N\}$. Attached to each link, an onramp and offramp are present.

A discretized version of the Lighthill–Whitham–Richards model [224, 249]

$$\partial_t u_\ell(t, x) + \partial_x f(u_\ell(t, x)) = 0, \tag{3.32}$$

Fig. 3.4 The flux function f, defined in (3.33)

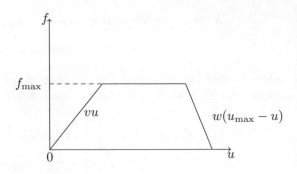

along each link I_ℓ is used. We recall that in (3.32) $u_\ell(t, x)$ represents the *density* of vehicles on link I_ℓ at time t and position x, and f gives the relationship between the density and the *flow* of vehicles, a relationship usually called *fundamental diagram*. Here we assume that the flux function $f : [0, u_{max}] \to \mathbb{R}$ has the following triangular form [100]:

$$f(u) = \min\{vu, w(u_{max} - u), f_{max}\}, \tag{3.33}$$

where v, w, u_{max}, and f_{max} are characteristics of the freeway; see Fig. 3.4.

The discrete model used here is inspired from those developed in [117, 247] and it is suitable for ramp metering applications. The discretization of (3.32) is composed by T time steps, N spatial cells or links, and N on-ramps and offramps. It is developed through a *Godunov-based* scheme [154, 206].

The state of cell $I_\ell \in \{1, \ldots, N\}$ at the numerical time $k \in \{1, \ldots, T\}$ is denoted by $u[\ell, k]$, while the number of vehicles on the adjacent onramp is given by $l[\ell, k]$. Moreover, we use the following additional functions:

- $\delta[\ell, k]$: Maximum flow of vehicles exiting link I_ℓ.
- $\sigma[\ell, k]$: Maximum flow of vehicles entering link I_ℓ.
- $d[\ell, k]$: Maximum flow of vehicles exiting onramp at link I_ℓ.
- r^{max}: Physical capacity of on-ramps.
- $f^{in}[\ell, k]$: Actual flow entering link I_ℓ.
- $f^{out}[\ell, k]$: Actual flow exiting link I_ℓ.
- $r[\ell, k]$: Actual flow exiting onramp I_ℓ.
- $\beta[\ell, k]$: Fraction of total flow from link I_ℓ entering link $I_{\ell+1}$ as opposed to offramp of link I_ℓ.
- p: Fraction of mainline flow given priority over onramp flow when merging in congestion.
- $D[\ell, k]$: Flow entering onramp of link I_ℓ.

Denoting with Δx and Δt respectively the space and time steps, the discrete system for (3.32) evolves from time k to $k + 1$ according to the following rules.

$$\delta[\ell, k] = \min\{vu[\ell, k], f_{max}\} \tag{3.34}$$

$$\sigma [\ell, k] = \min \left\{ w \left(u^{\max} - u [\ell, k] \right), f_{\max} \right\} \tag{3.35}$$

$$d [\ell, k] = \min \left\{ l [\ell, k] / \Delta t, r^{\max} \right\} \tag{3.36}$$

$$f^{\mathrm{in}} [\ell, k] = \min \left\{ \sigma [\ell, k], d [\ell - 1, k] + \beta [\ell, k] \delta [\ell, k] \right\} \tag{3.37}$$

$$f^{\mathrm{out}} [\ell, k] = \begin{cases} \delta [\ell, k] & \text{if } \dfrac{pf^{\mathrm{in}} [\ell + 1, k]}{\beta [\ell, k] (1 + p)} \geq \delta [\ell, k] \\[2ex] \dfrac{f^{\mathrm{in}} [\ell+1, k] - d [\ell+1, k]}{\beta [\ell, k]} & \text{if } \dfrac{f^{\mathrm{in}} [\ell + 1, k]}{1 + p} \geq d [\ell + 1, k] \\[2ex] \dfrac{pf^{\mathrm{in}} [\ell + 1, k]}{(1 + p) \beta [\ell, k]} & \text{otherwise} \end{cases} \tag{3.38}$$

$$r [\ell, k] = f^{\mathrm{in}} [\ell, k] - \beta [\ell, k] f^{\mathrm{out}} [\ell, k] \tag{3.39}$$

$$u [\ell, k + 1] = u [\ell, k] + \frac{\Delta t}{\Delta x} \left(f^{\mathrm{in}} [\ell, k] - f^{\mathrm{out}} [\ell, k] \right) \tag{3.40}$$

$$l [\ell, k + 1] = l [\ell, k] + \Delta t \left(D [\ell, k] - r [\ell, k] \right). \tag{3.41}$$

Equations (3.34)–(3.41) model the merging of onramp and mainline flows, as well as the propagation of congestion waves across the freeway network. The freeway-onramp-offramp junction shown in Fig. 3.5 gives a spatial relation of the state variables.

We also introduce a discrete control parameter $\omega [\ell, k] \in [0, 1]$, which represents a scaling factor on the demand of onramp related to I_ℓ at time step k. To this aim, we modify Eq. (3.36) in the following way

$$d [\ell, k] = \omega [\ell, k] \min \left\{ l [\ell, k] / \Delta t, r^{\max} \right\}. \tag{3.42}$$

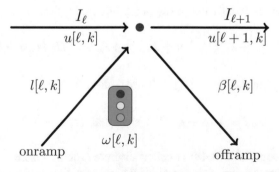

Fig. 3.5 A freeway-onramp-offramp junction. At time step k, the upstream mainline density $u [\ell, k]$ at link I_ℓ and onramp queue $l [\ell, k]$ merge and either exit the offramp with a split-ratio of $(1 - \beta)$ or continue onto the downstream mainline at link $I_{\ell+1}$. The control $\omega [\ell, k]$ scales the total demand from onramp $d [\ell, k]$ by a factor between 0 and 1

3.4.2   Optimal Control Problem

Using the model of Sect. 3.4.1, we develop a method to compute a control $\omega\,[\ell, k]$ for the ramp metering policy, over all the space indices $\ell \in \{1, \ldots, N\}$ and time $k \in \{1, \ldots, T\}$, which minimizes, or at least reduces, some specified objective.

More precisely, we consider the minimization of a function, which depends explicitly on the control variables ω and on the state variables u. Note that the state variables u depend on the control variables ω. It is possible to rewrite all the discrete Eqs. (3.34)–(3.35) in the compact way as

$$H\,(u, \omega) = 0. \tag{3.43}$$

Given some cost functional $J\,(u, \omega)$, the goal is to find an optimal control ω^* such that, denoting with u^* the corresponding optimal state, we have

$$J\left(u^*, \omega^*\right) = \min_{\omega} \, J\,(u, \omega) \tag{3.44}$$

$$H\left(u^*, \omega^*\right) = 0. \tag{3.45}$$

We compute the gradient of J with respect to the control variables ω, subject to the H constraints (3.45). Referring to the works [148, 160, 270], the generalized gradient of J is

$$\nabla_{\omega} J\left(u', \omega'\right) = \frac{\partial J\left(u', \omega'\right)}{\partial u} \frac{d\,u}{d\,\omega} + \frac{\partial J\left(u', \omega'\right)}{\partial \omega}. \tag{3.46}$$

By (3.45), the gradient of H with respect to ω is always zero and so

$$\nabla_{\omega} H = H_u d_{\omega} u + H_{\omega} = 0. \tag{3.47}$$

Adding last equation to (3.46) as a Lagrange-like multiplier:

$$\nabla_{\omega} J = J_u d_{\omega} u + J_{\omega} + \lambda^T \left(H_u d_{\omega} u + H_{\omega}\right) = \left(J_u + \lambda^T H_u\right) d_{\omega} u + \left(J_{\omega} + \lambda^T H_{\omega}\right).$$

Choosing λ such that $\left(J_u + \lambda^T H_u\right) = 0$, we deduce that

$$\nabla_{\omega} J = \left(J_{\omega} + \lambda^T H_{\omega}\right) \quad \text{such that: } H_u^T \lambda = -J_u. \tag{3.48}$$

The system of equations (3.48) is called discrete adjoint system, and we refer to [248] for a comprehensive description of the computation of the adjoint equations for ramp metering. Here we focus on an example focused on a specific application.

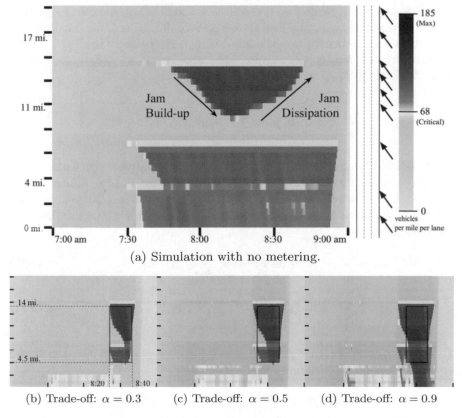

(a) Simulation with no metering.

(b) Trade-off: $\alpha = 0.3$ (c) Trade-off: $\alpha = 0.5$ (d) Trade-off: $\alpha = 0.9$

Fig. 3.6 (**a**) Depicts a space-time diagram of vehicle densities on 19.4 mile stretch of I15 Freeway with no ramp metering. The box objective, and example of *congestion-on-demand*, is applied in (**b**)–(**d**). The user specifies a "desired" traffic jam between postmile 4.5 and 14, for a duration of 20 minutes between 8:20 and 8:40. For this, the α parameter (introduced in Eq. (3.52)) enables the proper design of tradeoffs in the objective

3.4.3   Numerical Example

We will now apply the tools of adjoint-based finite-horizon optimal control and multi-objective optimization from the previous section to the case of coordinated ramp metering attacks. *Congestion-on-demand* describes a class of objectives where an attacker wishes to create congestion patterns of a specific nature. The attacks use a macroscopic freeway model of a 19.4 mile stretch of the I15 South Freeway in San Diego California. The model was split into 125 links with 9 on-ramps and was calibrated using loop-detector measurements available through the PeMS loop-detector system [182]. Figure 3.6a is a *Space-time diagram* of the I15 freeway. There is no ramp metering control applied to the simulation in Fig. 3.6b, i.e., the ramp meters are always set to green. In order to achieve the *congestion-on-demand* objective, we need to create a class of objective functions able to represent any jam pattern on the freeway. The method chosen is to maximize the traffic density where we want to put the congestion, while minimizing it everywhere else.

For every cell density value at position i and time k, we assign a coefficient $a_i^k \in \mathbb{R}$. We can then define the corresponding objective function:

$$J\left(\mathbf{u}, \boldsymbol{\rho}\right) = \sum_{i=1}^{N} \sum_{k=1}^{T} a_i^k \, \rho\left[i, k\right]. \tag{3.49}$$

Then, the box objective creates a box of congestion in the space-time diagram, i.e., congestion will be created on a specific segment of the freeway during a user-specified time interval. As there are two competing goals (maximize congestion in the box, minimize congestion elsewhere), the following two objective functions need to be considered:

$$f_1\left(\mathbf{u}, \boldsymbol{\rho}\right) = -\sum_{(i,k) \in \text{Box}} \rho\left[i, k\right], \tag{3.50}$$

$$\text{and } f_2\left(\mathbf{u}, \boldsymbol{\rho}\right) = \sum_{(i,k) \notin \text{Box}} \rho\left[i, k\right]. \tag{3.51}$$

To solve this multi-objective problem, a single parameter $\alpha \in [0, 1]$ is introduced and the following objective function is minimized:

$$J_\alpha\left(\mathbf{u}, \boldsymbol{\rho}\right) = \alpha \, f_1\left(\mathbf{u}, \boldsymbol{\rho}\right) + (1 - \alpha) \, f_2\left(\mathbf{u}, \boldsymbol{\rho}\right), \tag{3.52}$$

where α is a trade-off parameter: $\alpha = 1$ is complete priority on the congestion inside the box, while $\alpha = 0$ is complete priority on limiting density outside the box.

The results of the box objective are presented in Fig. 3.6b–d. The box of the objective is shown as a black frame with an actual size of 10 miles and 20 minutes. As the trade-off moves from $\alpha = 0.3$ to 0.9, there is a clear increase in the congestion within the box, at the expense of allowing the congestion to spill outside the desired bounds. In fact, Fig. 3.6d ($\alpha = 0.9$) activates the bottleneck near the top-left of the box earlier than Fig. 3.6b ($\alpha = 0.3$) to congest the middle portion of the box, which leads to a propagation of a congestion wave outside the bounds of the bottom-right of the box.

3.5   Optimal Control on Boundary and Flux Constraint

In this part we consider a stretch of road on which the controller can act on the upstream inflow and at a given point of the road by limiting the flow. We deal with the control problem

$$\begin{cases} \partial_t u + \partial_x f(u) = 0 & (t, x) \in \mathbb{R}^+ \times \mathbb{R}^+ \\ u(0, x) = u_0(x) & x \in \mathbb{R}^+ \\ f(u(t, 0)) = \omega_0(t) & t \in \mathbb{R}^+ \\ f(u(t, x_c)) \le \omega_c(t) & t \in \mathbb{R}^+, \end{cases} \tag{3.53}$$

where u_0 denotes the initial condition, ω_0 is the inflow control at $x = 0$, and ω_c is a time-dependent control imposing a flux constraint at a given position $x_c > 0$. The flux f is a function satisfying condition $(\mathcal{F})$. Solutions to (3.53) are to be intended according to the following definition.

Definition 14 A function $u \in \mathbf{C}^0 \left(\mathbb{R}^+; \mathbf{L}^\infty(\mathbb{R}^+; [0, 1]) \right)$ is a weak entropy solution to (3.53) if the following conditions hold.

1. For every test function $\varphi \in \mathbf{C}_c^1 \left(\mathbb{R}^2; \mathbb{R}^+ \right)$ and for every $k \in [0, 1]$

$$\int_0^{+\infty} \int_0^{+\infty} (|u - k| \, \partial_t \varphi + \text{sgn}(u - k) \, (f(u) - f(k)) \, \partial_x \varphi) \, \mathrm{d}x \, \mathrm{d}t$$

$$+ \int_0^{+\infty} \text{sgn} \left(f_L^{-1}(\omega_0(t)) - k \right) (f(u(t, 0+)) - f(k)) \, \varphi(t, 0) \, \mathrm{d}t$$

$$+ \int_0^{+\infty} |u_0 - k| \, \varphi(0, x) \, \mathrm{d}x + 2 \int_0^{+\infty} \left(1 - \frac{\omega_c(t)}{f(\sigma)} \right) f(k) \varphi(t, x_c) \, \mathrm{d}t \ge 0,$$

where f_L is defined in (3.3).
2. For a.e. $t \in \mathbb{R}^+$, $f(u(t, x_c-)) = f(u(t, x_c+)) \le \omega_c(t)$.

The above entropy condition is inherited from the entropy condition for conservation laws with space-discontinuous fluxes, see, e.g., [187, Definition 1.2]. For alternative equivalent definitions, see [17].

Remark 7 In Definition 14 we denote by $u(t, x_c\pm)$ the measure theoretic traces of u, which are implicitly defined by

$$\lim_{\varepsilon \to 0^+} \frac{1}{\varepsilon} \int_0^{+\infty} \int_{x_c}^{x_c+\varepsilon} |u(t, x) - u(t, x_c+)| \, \varphi(t, x) \, \mathrm{d}x \, \mathrm{d}t = 0$$

$$\lim_{\varepsilon \to 0^+} \frac{1}{\varepsilon} \int_0^{+\infty} \int_{x_c-\varepsilon}^{x_c} |u(t, x) - u(t, x_c-)| \, \varphi(t, x) \, \mathrm{d}x \, \mathrm{d}t = 0$$

for every $\varphi \in \mathbf{C}_c^1(\mathbb{R}^2; \mathbb{R})$. Note that both traces at x_c exist and are finite, by [16, Theorem 2.2].

Before stating the well-posedness result for (3.53), we need to introduce the domain

$$\mathcal{D} = \left\{ u \in \mathbf{L}^1 \left(\mathbb{R}^+; [0, 1] \right) : \text{sgn}(u - \sigma)(f(\sigma) - f(u)) \in \mathbf{BV}(\mathbb{R}^+; \mathbb{R}) \right\}. \tag{3.54}$$

Theorem 25 *Let $(\mathcal{F})$ holds. Fix $u_0 \in \mathcal{D}$ and $\omega_0, \omega_c \in \mathbf{BV}\left(\mathbb{R}^+; [0, f(\sigma)]\right)$. Then there exists a unique solution $u = u(t, x; u_0, \omega_0, \omega_c)$ to (3.53) in the sense of Definition 14 and, for every $t \in \mathbb{R}^+$, $u(t, \cdot; u_0, \omega_0, \omega_c) \in \mathcal{D}$.*

Moreover, if $u_0, u_0' \in \mathcal{D}$, $\omega_0, \omega_c, \omega_0', \omega_c' \in \mathbf{BV}\left(\mathbb{R}^+; [0, f(\sigma)]\right)$, then, for every $t > 0$, the following Lipschitz estimate on the corresponding solutions u, u' holds:

$$\left\| u(t) - u'(t) \right\|_{\mathbf{L}^1(\mathbb{R}^+; \mathbb{R})} \leq \left\| u_0 - u_0' \right\|_{\mathbf{L}^1(\mathbb{R}^+; \mathbb{R})} + \left\| \omega_0 - \omega_0' \right\|_{\mathbf{L}^1([0,t]; \mathbb{R})}$$
$$+ 2 \left\| \omega_c - \omega_c' \right\|_{\mathbf{L}^1([0,t]; \mathbb{R})}. \tag{3.55}$$

For a proof see [79].

3.5.1 Optimal Control Problems

In this part we deal with optimal control problems related to (3.53). More precisely, in the next paragraphs, we consider some cost functional, motivated by applications.

Queue Length Assume that at position x_c an obstacle reduces the traffic flow and consequently creates a queue propagating backwards. In this situation, it is reasonable that $\omega_c(t) \equiv \bar{\omega}_c$, i.e., the flow at x_c due to the presence of the obstacle is constant in time. Introduce, for every $u \in \mathcal{D}$, the congestion set

$$C_u = \{x \in [0, x_c[: \operatorname{sgn}(u(\xi+) - \sigma)(f(\sigma) - f(u)) = f(\sigma) - \bar{\omega}_c \text{ a.e. } \xi \in [x, x_c[\}. \tag{3.56}$$

The set $\mathcal{D}$ either is empty or is a real interval and represents the locations before the obstacle where the traffic flow is at the maximum level allowed by $\bar{\omega}_c$. The definition of the queue length functional J_{QL} is straightforward:

$$
\begin{aligned}
J_{QL} : \mathcal{D} &\longrightarrow && \mathbb{R}^+ \\
u &\longmapsto && \begin{cases} x_c - \inf C_u & \text{if } C_u \neq \emptyset \\ 0 & \text{if } C_u = \emptyset. \end{cases}
\end{aligned} \tag{3.57}
$$

As pointed out in [79, Proposition 2.5], the cost functional J_{QL} is upper semi-continuous with respect to the $\mathbf{L}^1$ topology, but not lower semicontinuous. The consequence is that there exists a control which maximizes the queue length. From the application point of view, it is reasonable to look for a control minimizing J_{QL}. Unfortunately, the regularity properties of J_{QL} are not sufficient to prove the existence of a minimizing control.

Stop and Go Waves The phenomenon of *stop & go waves* is well documented in engineering literature; see, for example, [189, 266]. The minimization of such waves is an important criterion for management of traffic.

The cost functional $J_{S\&G}$, for $T > 0$, is defined by

$$J_{S\&G} : \mathcal{D} \longrightarrow \mathbb{R}^+$$
$$u \longmapsto \int_0^T \int_0^{+\infty} p(x)\, \mathrm{d}\,|\partial_x\, v(u)|\, \mathrm{d}t, \tag{3.58}$$

where $p(x) \in [0, 1]$ is a given weight function, and $\int_0^{+\infty} p(x)\, \mathrm{d}\,|\partial_x\, v(u)|\, \mathrm{d}x$ measures the total variation of the velocity v. By assumption, v is a Lipschitz continuous function; this implies that the function $x \mapsto v(u(t, x))$ has finite total variation.

Proposition 6 *The functional $J_{S\&G} : \mathcal{D} \to \mathbb{R}^+$, defined in (3.58), is lower semicontinuous with respect to the $\mathbf{L}^1$ topology. Moreover, given $u_0 \in \mathcal{D}$ and $T > 0$, there exist $\omega_0^{opt}, \omega_c^{opt} \in \mathbf{BV}(\mathbb{R}^+; [0, f(\sigma)])$ such that*

$$J_{S\&G}(u^{opt}) = \min_{\omega_0, \omega_c \in \mathbf{BV}(\mathbb{R}^+; [0, f(\sigma)])} J_{S\&G}(u), \tag{3.59}$$

where u^{opt} denotes the solution to (3.53) with initial condition u_0 and controls $\omega_0^{opt}, \omega_c^{opt}$, while u is the solution to (3.53) with initial condition u_0 and controls ω_0, ω_c.

For a proof see [80, Lemma 2.1].

Travel Time The travel time represents a key quantity for drivers to be minimized. Assume that $x > x_c$ is the final destination for drivers starting their trip at $x = 0$. For simplicity, we also suppose that the initial condition $u_0 \equiv 0$. The total quantity of vehicles entering the road in the time interval $[0, T]$ ($T > 0$) is given by $Q_{in} = \int_0^T \omega_0(t)\, \mathrm{d}t$. Then one can consider the following cost functionals

$$J_{TT_1}(\omega_0, \omega_c) = \frac{1}{Q_{in}} \int_0^{+\infty} t f(u(t, \bar{x}))\, \mathrm{d}t \tag{3.60}$$

$$J_{TT_2}(\omega_0, \omega_c) = \frac{1}{Q_{in}} \int_0^{+\infty} t\, [f(u(t, \bar{x})) - f(u(t, 0))]\, \mathrm{d}t. \tag{3.61}$$

The following result holds. For a proof see [79].

Proposition 7 *The functionals J_{TT_1} and J_{TT_2}, defined in (3.60) and in (3.61), are Lipschitz continuous with respect to ω_0 and ω_c. Moreover, given $u_0 \equiv 0$, $T > 0$, and $\bar{x} > x_c$, there exist $\omega_0^{opt_1}, \omega_c^{opt_1}, \omega_0^{opt_2}, \omega_c^{opt_2} \in \mathbf{BV}(\mathbb{R}^+; [0, f(\sigma)])$ such that*

$$J_{TT_1}(u_0^{opt_1}, u_c^{opt_1}) = \min_{\omega_0, \omega_c \in \mathbf{BV}(\mathbb{R}^+; [0, f(\sigma)])} J_{TT_1}(u) \tag{3.62}$$

$$J_{TT_2}(u_0^{opt_2}, u_c^{opt_2}) = \min_{\omega_0, \omega_c \in \mathbf{BV}(\mathbb{R}^+; [0, f(\sigma)])} J_{TT_2}(u). \tag{3.63}$$

3.6 Optimization of Travel Time on Networks via Local Distribution Coefficients

In this section we show that optimizing the local traffic distribution coefficients, i.e., solving an optimization problem for the control system (3.1), allows us to find good sub-optimal solutions to the minimization of the travel time over large networks. The results contained in this section are from [64–66, 231].

The analytic treatment of an arbitrary network is beyond reach, due to the hybrid nature of the problem, having a continuous evolution and discrete set of parameters. We follow a strategy consisting of the following steps:

Step 1. Compute the optimal parameters for the asymptotic behavior of simple networks formed by a single junction.

Step 2. For a complex network, use the locally optimal parameters at every junction with a sample-and-hold technique: the current values of data at junctions are used and updated whenever they significantly change.

Step 3. Verify the performance of Step 2 comparing, via simulations, with other control strategies.

The first step is a hard task even for simple junctions; thus we focus on two special cases: the 2×1 case with two entering and one exiting road; and the 1×2 case with one entering and two exiting roads. For the first type of junction, one has only one right of way parameter q and we can solve the problem for different cost functionals. The second type of junctions has no right of way parameters and only one traffic distribution coefficient α. This case is even more complicate and so we address the issue only for a specific cost functional.

3.6.1 Optimization of Simple Networks

Consider a simple network with a single junction with n incoming roads I_i and m outgoing ones I_j. We start performing a heuristic computation of expected traffic load.

Denote by $c_{\varphi\psi}$, with $\varphi \in \{1, \ldots, n\}$ and $\psi \in \{n+1, \ldots, n+m\}$, the flux from source I_φ to destination I_ψ. Then, the following traffic load are expected on the roads: $u_\varphi = \sum_{\psi=n+1}^{n+m} c_{\varphi\psi}$ on road I_φ and $u_\psi = \sum_{\varphi=1}^{n} c_{\varphi\psi}$ on road I_ψ. Our strategy is as follows: use u_φ and u_ψ as initial data, solve the corresponding Riemann Problem at the junction, determining the density values $\widehat{u} = (\widehat{u}_1, \ldots, \widehat{u}_{n+m})$ at the junction, and use them as expected asymptotic values on the roads to optimize the expected travel times. Consider the following cost functions:

$$J_1 = \sum_{\varphi=1}^{n} \sum_{\psi=n+1}^{n+m} V_{\varphi\psi},$$

$$J_2 = \sum_{\varphi=1}^{n} \frac{1}{v_{\varphi}} + \sum_{\psi=n+1}^{n+m} \frac{1}{v_{\psi}},$$

$$J_3 = \sum_{\varphi=1}^{n} \sum_{\psi=n+1}^{n+m} c_{\varphi\psi} V_{\varphi\psi},$$

with $V_{\varphi\psi} = v_{\varphi} + v_{\psi}$, where $v_{\varphi} = v\left(\widehat{u}_{\varphi}\right)$ and $v_{\psi} = v\left(\widehat{u}_{\psi}\right)$ are the velocities on roads I_{φ} and I_{ψ}, respectively. Notice that J_1 gives the average speed over the network, J_2 the average travel time, while J_3 the averaged travel time weighted by the flow. We aim at maximizing J_1 and J_3, and at minimizing J_2 with respect to the traffic distribution parameters.

We start considering the case of $n = 2$ incoming roads and $m = 1$ outgoing ones, representing a merging. Then, the dynamics is determined by fixing a *right of way* parameter $q \in]0, 1[$; see Appendix B. To state our main results we need to introduce some notation: u_c is the critical density where the flux is maximized, γ_i^{max} indicates the maximal flux on road I_i, $q_1 = q$ and $q_2 = 1 - q$. For incoming roads we set:

$$s_{\varphi} = \begin{cases} -1 \text{ if } u_{\varphi} < u_c \text{ and } \gamma_1^{\max} + \gamma_2^{\max} \leq \gamma_3^{\max}, \\ \quad \text{or } u_{\varphi} < \sigma, \gamma_3^{\max} < \gamma_1^{\max} + \gamma_2^{\max}; \text{ and } q_{\varphi}\widehat{\gamma_3} \geq \gamma_{\varphi}^{\max}; \\ +1 \text{ if } u_{\varphi} \geq u_c, \\ \quad \text{or } u_{\varphi} < \sigma, \gamma_3^{\max} < \gamma_1^{\max} + \gamma_2^{\max}; \text{ and } q_{\varphi}\widehat{\gamma_3} < \gamma_{\varphi}^{\max}; \end{cases}$$

and for the outgoing one:

$$s_3 = \begin{cases} -1 \text{ if } u_3 \leq u_c, \\ \quad \text{or } u_3 > u_c, \gamma_1^{\max} + \gamma_2^{\max} < \gamma_3^{\max}; \\ +1 \text{ if } u_3 > u_c, \gamma_1^{\max} + \gamma_2^{\max} \geq \gamma_3^{\max}. \end{cases}$$

For the simple case of the merge, the functional J_3 does not depend on the choice of the right of way parameter q, while we can explicitly find the optimal ones for the functionals J_1 and J_2, (see [64]):

Theorem 26 *Consider a junction J with $n = 1$ incoming road and $m = 2$ outgoing roads and define: $k^- = \dfrac{f(\widehat{u}_c) - \gamma_1^{max}}{\gamma_1^{max}}$, $k^+ = \dfrac{\gamma_2^{max}}{f(\widehat{u}_c) - \gamma_2^{max}}$. The cost functions J_1 and J_2 are maximized for the same values of q, given by*

1. for $s_1 = s_2 = +1$:

(a) $q \in \left[0, \dfrac{1}{1+k^+}\right]$, if $k^- \leq 1 \leq k^+$, with $k^- k^+ > 1$, or $1 \leq k^- \leq k^+$;

(b) $q \in \left[0, \frac{1}{1+k^+}\right] \cup \left]\frac{1}{1+k^-}, 1\right]$, if $k^- \leq 1 \leq k^+$, with $k^-k^+ = 1$;

(c) $q \in \left[\frac{1}{1+k^-}, 1\right]$, if $k^- \leq 1 \leq k^+$, with $k^-k^+ < 1$, or $k^- \leq k^+ \leq 1$;

2. for $s_1 = s_2 = -1$:

(a) $q \in \left[0, \frac{1}{1+k^+}\right]$, if $k^- \leq 1 \leq k^+$, with $k^-k^+ < 1$, or $k^- \leq k^+ \leq 1$;

(b) $q \in \left[0, \frac{1}{1+k^+}\right] \cup \left]\frac{1}{1+k^-}, 1\right]$, if $k^- \leq 1 \leq k^+$, with $k^-k^+ = 1$;

(c) $q \in \left[\frac{1}{1+k^-}, 1\right]$, if $k^- \leq 1 \leq k^+$, with $k^-k^+ > 1$, or $1 \leq k^- \leq k^+$;

3. for $s_1 = -1 = -s_2$: $q \in \left[\frac{1}{1+k^-}, 1\right]$;

4. for $s_1 = +1 = -s_2$: $q \in \left[0, \frac{1}{1+k^+}\right]$.

Let us now pass to the case of $n = 1$ incoming road and $m = 2$ outgoing roads, representing a bifurcation. Let α denote the fraction of traffic flowing from road I_1 to road I_2 (while $1 - \alpha$ is the fraction flowing to road I_3). The optimization of the functionals J_1 and J_2 are similar to the case of merge, while for J_3 we have the following (see [64]):

Theorem 27 *Define:*

$$\bar{\alpha} = \frac{\gamma_b^{\max}}{\gamma_b^{\max} + \gamma_c^{\max}}.$$

If

$$\gamma_a^{\max} \leq \min\left\{\frac{\gamma_b^{\max}}{\bar{\alpha}}, \frac{\gamma_c^{\max}}{1 - \bar{\alpha}}\right\}, \tag{3.64}$$

then the maximal value for J_3 is obtained for all the values of α so that (3.64) holds true with $\bar{\alpha}$ replaced by α. In the opposite case, J_3 is maximized only for $\alpha = \bar{\alpha}$.

3.6.2 Simulations of Two Urban Networks

In this section we use the results stated in Section 3.6.1 to optimize traffic on two urban networks: the first one is a large traffic circle in Rome (Italy), while the second is a small network located in the urban area of the city of Salerno (Italy). More precisely, our first network represents the large traffic circle of the *Re di Roma Square*. Such network experiences high traffic and congestion every day, also because of the heavy traffic load in the morning rush hours. The square becomes completely stuck with large delays for users. The second network is given by a junction on *Via Parmenide*, also experiencing high traffic also due to a particularly

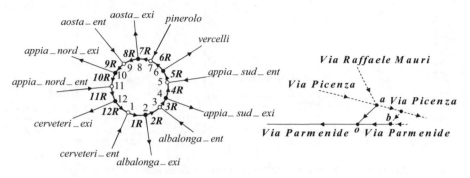

Fig. 3.7 Topology of Re di Roma Square (left) and Via Parmenide crossing in Salerno (right)

long red cycle for an incoming road. The result illustrated in this section were developed in [65].

Our aim is to show the effectiveness of the approach also by comparison with other control algorithms, including random ones. Beside the cost functions introduced in Sect. 3.6.1 we will consider stop and go waves so the functional $J_{S\&G}$ introduced in (3.58). We consider the case $p(x) \equiv 1$ and the flux $f(u) = u(1 - u)$ thus we can rewrite:

$$J_{S\&G} = \int_0^T \int_{\cup I_i} |Dv(u)| \, dt \, dx,$$

indeed the variation of u is equivalent to that of $v(u) = 1 - u$. This cost function provides also a measure of safety, since velocity differences are responsible for many accidents.

The *Re di Roma Square* network consists of junctions with two incoming and one outgoing roads (2×1 junctions) and junctions with one incoming and two outgoing roads (1×2 junctions). For the latter we consider distribution coefficient given by the road maximal capacities, while we optimize over right of way parameters of the latter. Figure 3.7 (left) illustrate the topology of the network: 2×1 junctions (1, 3, 5, 7, 9, 11) are in white, 1×2 junctions (2, 4, 6, 8, 10, 12) are in black.

The second network consists of a small area of the Salerno urban network, see Fig. 3.7 (right). We are particularly interested in the signalized junction, denoted o, with two incoming roads and one outgoing road. One incoming road, i.e., $a - o$, is very short and connects Via Picenza to Via Parmenide. The traffic light cycle is of two minutes, with green phase of only 15 seconds for incoming road $a - o$. Accordingly we set the right of way parameter as follows:

$$p = \frac{105}{120} = 0.875,$$

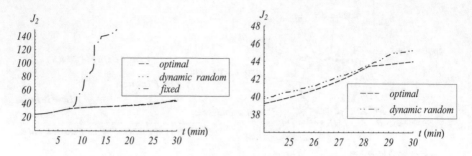

Fig. 3.8 Time evolution of cost J_2 for different control policies (left) and zoom (right)

for road $b - o$, while for road $a - o$

$$q = 1 - p = 1 - 0.875 = 0.125.$$

Simulations are performed over a time interval of $T = 30$ minutes, with flux function given by $f(u) = u(1 - u)$. At the initial time, roads are empty, thus the density is set to zero. For the first network, we consider the inflow values of 0.3 and 0.75, while for the second network the inflow value is 0.8 for roads entering junction o and 0.3 for the outgoing ones. We compare three cases: (1) right of way parameters optimizing the cost functionals J_1 and J_2 called *optimal case*; fixed right of way parameters, called *fixed case*) choosing $p = 0.2$ for first network and $p = 0.875$ for the second; dynamic random parameters, called *dynamic random case*, with parameters randomly sampled with uniform distribution at every time step. Figure 3.8 illustrates the time evolution of the cost functional J_2 for the first network and the different choices of the right of way parameters. Such choices reflect different traffic control policies. The fixed control policy is very poorly performing, while the other two are comparable with the optimal one slightly preferable as shown in the zoomed area Fig. 3.8 (right). Even if the two policies, optimal and random, perform similarly, we can capture the difference in traffic patterns by looking at the functional $J_{S\&G}$, representing the smoothness of traffic, see Fig. 3.9. The optimal policy significantly outperforms the others. On the other side, the dynamic random choice may generate high oscillations, which in turn compromise safety. For the second network, we consider the cost function J_1, whose time evolution is depicted in Fig. 3.10. The optimal control policy outperforms the others with an advantage of around 20% in terms of J_1 values. For the dynamic random choice J_1 converges numerically to the value corresponding to the fixed parameters $p = 0.5$ (see [65]). This can be understood as follows. Since the flux is given by $f(\rho) = \rho(1 - \rho)$ we have:

$$J_1(p) = \chi - \frac{1}{2}\sqrt{1 - 4cp} - \frac{1}{2}\sqrt{1 - 4c(1 - p)},$$

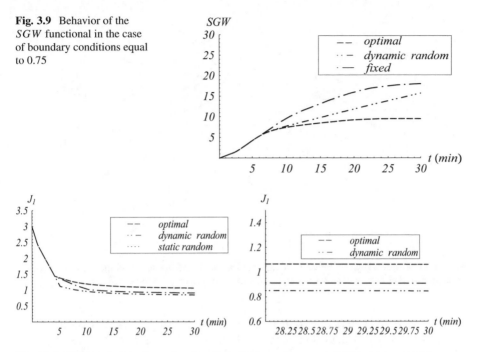

Fig. 3.9 Behavior of the SGW functional in the case of boundary conditions equal to 0.75

Fig. 3.10 Time evolution of cost function J_1 for the second network (left) and zoom (right)

where χ and $c > 0$ are constant, not depending on p. Therefore, $\frac{dJ_1(p)}{dp} \geq 0$ if and only if $p \in [0.5, 1]$. In other words, the dynamic random choice approach the worst choice in terms of J_1.

3.6.3   Emergency Management

We now turn to another optimization problem by focusing on emergency situation. The results in this section were developed in [231].

In heavy loaded network it is important to create safe corridors for emergency vehicles, see Fig. 3.11 for a pictorial presentation of the problem. More precisely, we consider a network made of junctions, each with two incoming and two outgoing roads. Then the dynamics is determined by the traffic distribution coefficients, called α and β (see Appendix B). Naming incoming roads as I_1 and I_2 and outgoing ones I_3 and I_4, α is the percentage of traffic from road I_1 going to road I_3 and similarly β is the percentage of traffic from road I_2 going to road I_3.

The emergency vehicles are assumed to follow a velocity function given by:

$$\omega(u) = 1 - \delta + \delta v(u),   \tag{3.65}$$

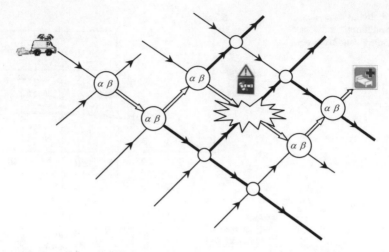

Fig. 3.11 Cartoon representation of a car accident on a road network and the creation of a safe corridor for emergency vehicles via regulation of the traffic distribution coefficients

with $0 < \delta < 1$, $v(u) = u(1 - u)$ being the velocity of regular traffic. Since $\omega(u_{\max}) = 1 - \delta > 0$, the emergency vehicles travel faster than other vehicles.

The general problem we aim to solve is the following. Given a junction with n incoming roads, say I_φ, $\varphi = 1, \ldots, n,$, and m outgoing ones, say I_ψ, $\psi = n + 1, \ldots, n + m$, and initial data $(u_{\varphi,0}, u_{\psi,0})$, we aim to maximize the cost:

$$W_{\varphi,\psi}(t) = \int_{I_\varphi} \omega\left(u_\varphi(t, x)\right) \mathrm{d}x + \int_{I_\psi} \omega\left(u_\psi(t, x)\right) \mathrm{d}x.$$

which gives the traveling time of emergency vehicles on the specific corridor formed by incoming road I_φ and outgoing one I_ψ. As above, we consider the asymptotic state of the Riemann Problem to solve the optimization.

Focusing on 2×2 junctions, with $I_{1,2}$ incoming roads and $I_{3,4}$ outgoing ones, for sufficiently big time T we get:

$$W_{\varphi,\psi}(T) = \omega\left(\widehat{u}_\varphi\right) + \omega\left(\widehat{u}_\psi\right) = 2 - \delta - \frac{\delta}{2}\left(s_\varphi\sqrt{1 - 4\widehat{\gamma}_\varphi} + s_\psi\sqrt{1 - 4\widehat{\gamma}_\psi}\right),$$

(3.66)

where s_φ and s_ψ are defined as:

$$s_\varphi = \begin{cases} +1, & \text{if } u_{\varphi,0} \geq \frac{1}{2}, \text{ or } u_{\varphi,0} < \frac{1}{2} \text{ and } \gamma_\varphi^{\max} > \widehat{\gamma}_\varphi, \\ -1 & \text{if } u_{\varphi,0} < \frac{1}{2} \text{ and } \gamma_\varphi^{\max} = \widehat{\gamma}_\varphi, \end{cases}$$

$$s_\psi = \begin{cases} +1, & \text{if } u_{\psi,0} > \frac{1}{2} \text{ and } \gamma_\psi^{\max} = \widehat{\gamma}_\psi, \\ -1 & \text{if } u_{\psi,0} \leq \frac{1}{2}, \text{ or } u_{\psi,0} > \frac{1}{2} \text{ and } \gamma_\psi^{\max} > \widehat{\gamma}_\psi. \end{cases}$$

Table 3.1 Initial conditions
for the three simulation cases

	$u_{a,0}$	$u_{b,0}$	$u_{c,0}$	$u_{d,0}$
Case A	0.15	0.6	0.8	0.9
Case B	0.15	0.6	0.9	0.8
Case C	0.25	0.1	0.85	0.95

Fixing, without loss of generality, the safe corridor given by $\varphi = 1$ and $\psi = 3$, we get:

$$W_{1,3}(T) = \omega(\widehat{u}_1) + \omega(\widehat{u}_3) = 2 - \delta - \frac{\delta}{2}\left(s_1\sqrt{1 - 4\widehat{\gamma}_1} + s_3\sqrt{1 - 4\widehat{\gamma}_3}\right). \quad (3.67)$$

We can find optimal values of the traffic distribution coefficients (see [231]):

Theorem 28 *Consider a junction with incoming roads, I_1 and I_2 and outgoing ones I_3 and I_4. For T sufficiently big, the cost $W_{1,3}(T)$ is maximized by the traffic distribution coefficients $\alpha_{opt} = 1 - \frac{\gamma_d^{\max}}{\gamma_a^{\max}}$ and any $\beta_{opt} \in \left[0, 1 - \frac{\gamma_d^{\max}}{\gamma_a^{\max}}\right]$, with the exception of the following two cases.*

If $\gamma_1^{\max} \leq \gamma_4^{\max}$ then there is no optimal value, but the minimum is approximated by $\alpha = \varepsilon_1$ and $\beta_{opt} = \varepsilon_2$, for ε_1 and ε_2 small, positive and such that $\varepsilon_1 \neq \varepsilon_2$.

Similarly, if $\gamma_1^{\max} > \gamma_3^{\max} + \gamma_4^{\max}$, then there is no optimal value, but the minimum is approximated by $\alpha = \frac{\gamma_3^{\max}}{\gamma_3^{\max} + \gamma_4^{\max}} - \varepsilon_1$ and $\beta = \frac{\gamma_3^{\max}}{\gamma_3^{\max} + \gamma_4^{\max}} - \varepsilon_2$, with $\varepsilon_{1,2}$ as for the first case.

To test the effectiveness of the control prescribed by Theorem 28, we compute the cost function evolution and compare with random coefficients, i.e., parameters taken randomly when the simulation starts and then kept constant. More precisely, we consider three scenarios denoted by A, B, and C, with initial data reported in Table 3.1 and boundary conditions coinciding with initial data. The values prescribed by Theorem 28 are as follows: for case A, $\alpha_{opt} = 0.294118$ and $0 \leq \beta_{opt} < \alpha_{opt}$ (we choose β_{opt} equal to 0.2); for case B, $\alpha_{opt} = \varepsilon_1$, $\beta_{opt} = \varepsilon_2$; for case C, $\alpha_{opt} = 0.708571 + \varepsilon_1$, $\beta_{opt} = 0.708571 + \varepsilon_2$. Figures 3.12, 3.13, and 3.14 show the time evolution of the cost function $W_{a,c}(t)$ and the values of $W_{a,c}(T)$ as function of the two parameters α and β in cases A, B, and C, respectively, with $\delta = 0.5$ and $T = 30$ minutes. The simulations confirm the optimality of the controls prescribed by Theorem 28. We also notice, especially in Fig. 3.14, that the optimal control may not achieve the best value for transient times, but do so for sufficiently large times.

3.7 Bibliographical Notes

The possibility of controlling the solutions to conservation laws on networks, by acting on the Riemann solvers through control parameters, has been addressed

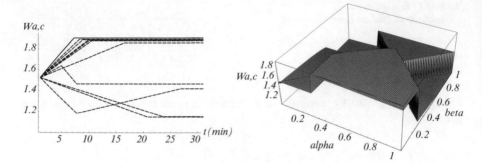

Fig. 3.12 Case A, evolution of $W_{a,c}(t)$; left: choice of optimal distribution coefficients (continuous line) and random parameters (dashed lines); right: 3D plots of $W_{a,c}(T)$

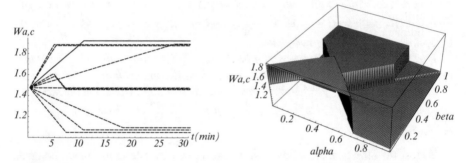

Fig. 3.13 Case B, evolution of $W_{a,c}(t)$; left: choice of optimal distribution coefficients (continuous line) and random parameters (dashed lines); right: 3D plots of $W_{a,c}(T)$

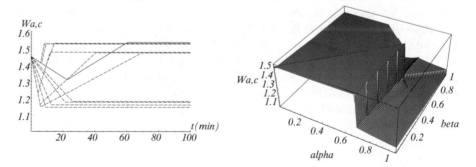

Fig. 3.14 Case C, evolution of $W_{a,c}(t)$; left: choice of optimal distribution coefficients (continuous line) and random parameters (dashed lines); right: 3D plots of $W_{a,c}(T)$

first in [144] in 2009. A different approach, based on solutions to boundary value problems, has been introduced in [10] in 2018. In the context of road traffic, the setting in [10] fits well with the minimization of travel time or of other meaningful functionals.

The modeling and the (optimal) control of signalized intersections have been widely studied in literature, both from the mathematical [70] and engineering [147, 250, 253, 276] point of view. The approach described in Section 3.3 is based on the papers [165, 166]; see also [139, Chapter 7.1].

In the case of freeways, ramp metering is a widely used strategy for controlling traffic flow without acting directly on the mainline flow or using toll systems. This strategy has been used mostly in connection with microscopic traffic models; see, for example, [37]. The use of feedback controls, in order to obtain a desired flow in the main line, has been studied in various papers [2, 237, 254]. The predictive metering strategy instead uses a model, possibly with uncertainties, for predicting the inflow over a finite time horizon, and consequently it deduces various policies for controlling the flow; see [121, 155, 194]. Section 3.4, based on the paper [247], uses the adjoint calculus for finding solutions to optimal control problem; see also [148, 160, 270, 271].

Conservation laws with flux constraints have been introduced in [78, 140] in 2007. In [78] the problem was presented by a control perspective and a well-posedness result was proved. In [140] the model was proposed for describing the situation of a bottleneck on a road. Various Riemann solvers have been introduced together with existence and uniqueness results. Generalizations on (possibly moving) flux constraints both for scalar equations and for systems were considered in various papers; see [111–114, 137, 138, 145, 272].

For the optimization of travel time or other significant functional on networks via the choice of distribution of coefficients, see [64, 66, 231].

Chapter 4
Distributed Control for Conservation Laws

4.1 Introduction

This chapter focuses on control of systems of conservation laws with distributed parameters. Problem with different parameterized fluxes is addressed: in particular, we deal with cases where the control is the maximal speed and look for continuous dependence of the solution on parameters. Various examples of conservation laws with distributed control are presented. The control, in this case, acts on the flux or on the parameters of the flux, as for example the maximal speed.

In Sect. 4.2, we introduce the classical notions of Riemann Solver Semigroup solutions [51] and its stability [39].

Subsequently, in Sect. 4.3, we consider the case of variable speed limit computed via needle-like variation methods. Specifically, an optimal control problem for traffic flow on a single road using a variable speed limit is studied. The control variable is the maximal allowed velocity, which is a function of time with finite total variation, and the aim is to obtain a prescribed outgoing flow. More precisely, the main goal is to minimize the quadratic difference between the achieved outflow and the given target outflow. Mathematically, the problem is very hard because of the delays in the effect of the control variable (speed limit). In Sects. 4.4 and 4.5, following respectively [153] and [193], the problem of variable speed limit is analyzed by discrete-optimization methods for first and second order traffic models on networks. To set up the optimization problem, a cost functional is identified, while the constraints are given by the discretized LWR or Aw–Rascle–Zhang models.

© The Author(s) 2022
A. Bayen et al., *Control Problems for Conservation Laws with Traffic Applications*,
PNLDE Subseries in Control 99, https://doi.org/10.1007/978-3-030-93015-8_4

4.2   Riemann Solver Semigroup and Stability

This section deals with the notion of Standard Riemann Semigroup (SRS) for a strictly hyperbolic system of conservation laws in one space dimension and with the dependence of the solution on the flux.

4.2.1   Classical Riemann Solver Semigroup Solutions

Here we focus the attention on the following Cauchy problem for a strictly hyperbolic $n \times n$ system of conservation laws in one space dimension:

$$\begin{cases} \partial_t u + \partial_x f(u) = 0 \\ u(0, x) = u_0(x). \end{cases} \tag{4.1}$$

For initial data $u_0 \in \mathbf{L}^1(\mathbb{R}; \mathbb{R}^n)$ with small total variation, Glimm's theorem [152] provides the global existence of weak solutions. The well-posedness of (4.1) is contained in the following result, due to Bressan [54].

Theorem 29 *Let $\Omega \subseteq \mathbb{R}^n$ be an open set containing the origin, and let $f : \Omega \to \mathbb{R}^n$ be a smooth map. Assume that the system (4.1) is strictly hyperbolic and that each characteristic field is either linearly degenerate or genuinely nonlinear. Then there exist a closed domain $\mathcal{D} \subset \mathbf{L}^1(\mathbb{R}; \mathbb{R}^n)$, positive constants η_0 and L, and a continuous semigroup $S : [0, +\infty[\times \mathcal{D} \to \mathcal{D}$ with the following properties:*

(i) *Every function $u_0 \in \mathbf{L}^1(\mathbb{R}; \mathbb{R}^n)$ with $\mathrm{TV}(u_0) \leq \eta_0$ belongs to $\mathcal{D}$.*
(ii) *For all $u_0, v_0 \in \mathcal{D}$, $t, s \geq 0$, one has*

$$\|S_t u_0 - S_s v_0\|_{\mathbf{L}^1} \leq L \left(|t - s| + \|u_0 - v_0\|_{\mathbf{L}^1} \right).$$

(iii) *If $u_0 \in \mathcal{D}$ is a piecewise constant function, then for $t > 0$ sufficiently small the function $u(t, \cdot) = S_t u_0$ coincides with the solution of (4.1), obtained by piecing together the standard self-similar solutions of the corresponding Riemann problems.*

The invariant domain $\mathcal{D}$ in Theorem 29 can be chosen in the form

$$\mathcal{D} = \mathrm{cl} \left\{ u \in \mathbf{L}^1(\mathbb{R}; \mathbb{R}^n) : u \text{ piecewise constant, } V(u) + C \cdot Q(u) < \delta_0 \right\}, \tag{4.2}$$

for suitable positive constants C and δ_0. Here $V(u)$ and $Q(u)$ denote the total strength of waves and the wave interaction potential of u, while "cl" denotes the closure in $\mathbf{L}^1$; see [51]. Following [50], we say that a map S with the properties (i), (ii), and (iii) of Theorem 29 is a *Standard Riemann Semigroup* (SRS). We remark that Theorem 29 is also valid in case of general 2×2 systems [53] and of special systems with coinciding shock and rarefaction curves [48, 49]. Moreover, solutions can be obtained by viscous approximations as shown in the seminal paper [38].

The next result deals with the uniqueness of an SRS; for the proof, see [50].

Theorem 30 *For a given domain $\mathcal{D}$ of the form (4.2), there is at most one continuous semigroup $S : [0, +\infty[\times \mathcal{D} \to \mathcal{D}$ satisfying conditions (i), (ii), and (iii) in Theorem 29. Moreover, if an SRS does exist, then the following properties hold:*

(iv) *Each trajectory $t \mapsto u(t, \cdot) := S_t u_0$ is a weak entropy admissible solution of the corresponding Cauchy problem (4.1).*

(v) *Let $(u_\nu)_{\nu \geq 1}$ be a sequence of approximate solutions of (4.1), generated by a wave-front tracking algorithm or by the Glimm scheme with uniformly distributed sampling. Then, for every $t > 0$,*

$$\lim_{\nu \to +\infty} \|u_\nu(t) - S_t u_0\|_{\mathbf{L}^1} = 0.$$

(vi) *Let $u = u(t, x)$ be a piecewise Lipschitz continuous entropy admissible solution of (4.1) defined on $[0, T] \times \mathbb{R}$ for some $T > 0$. Then $u(t, \cdot) = S_t u_0$ for all $t \in [0, T]$.*

4.2.2   Stability of the Standard Riemann Semigroup

In this part, we consider the dependence of the solution of Riemann problems with respect to the flux function f. According to Theorem 29, if f is a smooth function and the system

$$\partial_t u + \partial_x f(u) = 0 \qquad (4.3)$$

is strictly hyperbolic with each characteristic field either linear degenerate of genuinely nonlinear, then there exist a domain $\mathcal{D}^f$ and an SRS $S^f : [0, +\infty[\times \mathcal{D}^f \to \mathcal{D}^f$. We define by $\mathbf{Hyp}(\Omega)$ the set containing all the fluxes satisfying the previous assumptions.

We introduce now a concept similar to a metric between fluxes in $\mathbf{Hyp}(\Omega)$. To this aim, for every $f \in \mathbf{Hyp}(\Omega)$, define the set

$$\mathcal{R}^f = \left\{ (u^-, u^+) \in \Omega \times \Omega : u^- \neq u^+, \ u^- \chi_{(-1,0)} + u^+ \chi_{(0,1)} \in \mathcal{D}^f \right\},$$

where χ denotes the characteristic function of a set. Note that, if $(u^-, u^+) \in \mathcal{R}^f$, then the Riemann problem

$$\begin{cases} \partial_t u + \partial_x f(u) = 0 \\ u(0, x) = \begin{cases} u^- & \text{if } x < 0 \\ u^+ & \text{if } x > 0 \end{cases} \end{cases}$$

admits an entropy admissible solution. Let f_1, $f_2 \in \mathbf{Hyp}(\Omega)$ with $\mathcal{D}^{f_2} \subseteq \mathcal{D}^{f_1}$, and let us define the "distance" between f_1 and f_2 as

$$\hat{d}(f_1, f_2) = \sup_{(u^-, u^+) \in \mathcal{R}^{f_2}} \frac{1}{|u^+ - u^-|} \cdot \left\| S_1^{f_1}\left(\tilde{u}_{(u^-, u^+)}\right) - S_1^{f_2}\left(\tilde{u}_{(u^-, u^+)}\right) \right\|_{\mathbf{L}^1}, \tag{4.4}$$

where $\tilde{u}_{(u^-, u^+)} = u^- \chi_{(-\infty, 0)} + u^+ \chi_{(0, +\infty)}$.

The next result, whose proof is contained in [39], is the stability estimate with respect to the flux functions.

Theorem 31 *Let $f_1 \in \mathbf{Hyp}(\Omega)$. Then there exists a positive constant L_{f_1} such that, for every $f_2 \in \mathbf{Hyp}(\Omega)$ with $\mathcal{D}^{f_2} \subseteq \mathcal{D}^{f_1}$, for all $u \in \mathcal{D}^{f_2}$, and for all $t > 0$, it holds*

$$\left\| S_t^{f_1} u - S_t^{f_2} u \right\|_{\mathbf{L}^1} \leq L_{f_1} \hat{d}(f_1, f_2) \int_0^t \mathrm{TV}\left(S_t^{f_2} u\right) \mathrm{d}t. \tag{4.5}$$

4.3   Needle-Like Variations for Variable Speed Limit

This section is devoted to the specific problem of controlling traffic via variable speed limit. The main idea is the use of needle-like variations to compute the control policy as used in the classical Pontryagin Maximum Principle [55]. We study the following Initial Boundary Value Problem (IBVP):

$$\begin{cases} u_t + f(u, v(t))_x = 0, & (t, x) \in \mathbb{R}^+ \times [0, L], \\ u(0, x) = u_0(x), & x \in [0, L], \\ f_l(t) = \mathrm{In}(t), \\ f_r(t) = u(t, L)\, v(t), \end{cases} \tag{4.6}$$

where u_0 is the initial condition, f_l (f_r) is the left (right) boundary flow, In is the time-dependent inflow, and $v(t)$ is the maximal speed and the control variable, see Fig. 4.1a. It takes value on a bounded interval $v(t) \in [v_{\min}, v_{\max}]$. The flux function $f : [0, u_{\max}] \times [v_{\min}, v_{\max}] \to \mathbb{R}^+$ is given by

$$f(u, v(t)) = \begin{cases} u v(t), & \text{if } 0 \leq u \leq u_{\mathrm{cr}}, \\ \dfrac{v(t) u_{\mathrm{cr}}}{u_{\max} - u_{\mathrm{cr}}} (u_{\max} - u), & \text{if } u_{\mathrm{cr}} < u \leq u_{\max}, \end{cases} \tag{4.7}$$

see Fig. 4.1b.

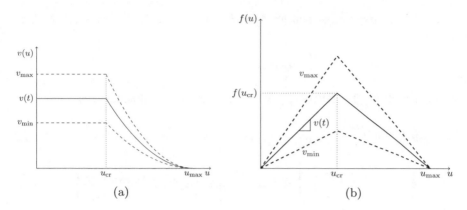

Fig. 4.1 Velocity and flow for different speed limits. (**a**) Velocity function. (**b**) Newell–Daganzo fundamental diagram

4.3.1 Variable Speed Limit: Control Problem

In this section, we introduce the mathematical framework for the speed regulation problem. Given an inflow $In(t)$, we want to track a fixed outflow $Out(t)$ on a time horizon $[0, T]$, $T > 0$, by acting on the time-dependent maximal velocity $v(t)$. A maximal velocity function $v : [0, T] \to [v_{\min}, v_{\max}]$ is called a **control policy**.

It is easy to see that a road in free flow can become congested only because of the outflow regulation with shocks moving backward, see [57, Lemma 2.3]. Since we assume Neumann boundary conditions at the road exit, the traffic will always remain in free flow, i.e., $u(t, x) \leq u_{\mathrm{cr}}$ for every $(t, x) \in [0, T] \times [0, L]$. Given the inflow function $In(t)$, we consider the Initial Boundary Value Problem (4.6) with assigned flow boundary condition $f_l \doteq f(u(t, 0^+))$ on the left in the sense of Bardos, Le Roux, and Nedelec, see [29], and Neumann boundary condition (flow $f_r \doteq f(u(t, 0^-)))$ on the right. Let us make the following assumptions:

Hypothesis 1 *There exist* $0 < u_0^{\min} \leq u_0^{\max} \leq u_{\mathrm{cr}}$ *and* $0 < f^{\min} \leq f^{\max}$ *such that* $u_0 \in \mathbf{BV}([0, L], [u_0^{\min}, u_0^{\max}])$ *and* $In \in \mathbf{BV}([0, T], [f^{\min}, f^{\max}])$.

Hypothesis 2 *We assume Hypothesis 1 and the following:*

$$u_0^{\min} \leq \frac{f^{\min}}{v_{\max}} \qquad and \qquad u_0^{\max} \geq \frac{f^{\max}}{v_{\min}}.$$

Hypothesis 1 gives directly the following proposition:

Proposition 8 *Assume that Hypothesis 1 holds and*

$$v \in \mathbf{BV}([0, T], [v_{\min}, v_{\max}]).$$

Then, there exists a unique entropy solution $u(t, x)$ to (4.6). Moreover, $u(t, x) \leq u_{cr}$ and, setting

$$Out(t) = u(t, L)v(t), \tag{4.8}$$

we have that $Out(.) \in \mathbf{BV}([0, T], \mathbb{R})$ and the following estimates hold:

$$\min\left\{u_0^{\min}, \frac{f^{\min}}{v_{\max}}\right\} \leq u(t, x) \leq \max\left\{u_0^{\max}, \frac{f^{\max}}{v_{\min}}\right\}, \ for \ x \in [0, L] \tag{4.9}$$

$$\min\left\{u_0^{\min} v_{\min}, f^{\min}\frac{v_{\min}}{v_{\max}}\right\} \leq Out(t) \leq \max\left\{u_0^{\max} v_{\max}, f^{\max}\frac{v_{\max}}{v_{\min}}\right\}. \tag{4.10}$$

For the proof, we refer the reader to [116].

Definition 15 A Link Entering Time (LET) function $\tau = \tau(t, v)$ is defined as the entering time for a car exiting the road at time t given a control policy v.
For every t_0 satisfying $\int_0^{t_0} v(s)ds = L$ and for every $t \geq t_0$, we get

$$\int_{\tau(t)}^{t} v(s)ds = L. \tag{4.11}$$

Such $\tau(t)$ is unique, due to the hypothesis $v \geq v_{\min} > 0$.

Remark 8 The function depends on the control policy v, but for simplicity we will write $\tau(t)$ when the policy is clear from the context. Notice that LET is defined only for time greater than a given $t_0 > 0$, the exit time of the car entering the road at time $t = 0$, see Fig. 4.2.

Fig. 4.2 Graphical representation of the LET function $\tau = \tau(t, v)$ defined in (4.11)

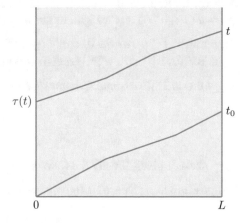

From the identity

$$\int_{\tau(t_1)}^{\tau(t_2)} v(s)ds = \int_{t_1}^{t_2} v(s)ds,$$

we get the following lemma:

Lemma 3 *Given a control policy v, the function τ is a Lipschitz continuous function, with Lipschitz constant $\dfrac{v_{max}}{v_{min}}$.*

Recalling the definition of outflow of the solution given in (4.8), we get the following proposition:

Proposition 9 *The input–output flow map of the Initial Boundary Value Problem (briefly IBVP) (4.6) is given by*

$$Out(t) = In(\tau(t))\frac{v(t)}{v(\tau(t))}. \tag{4.12}$$

For the proof, we refer the reader to [116]. Classical techniques of linear control cannot be used since the map (4.28) is highly nonlinear with respect to the control v. In fact, the effect of the control v at time t on the outflow depends on the choice of v on the time interval $[\tau(t), t]$ because of the presence of the LET map in formula (4.12). This clearly shows how delays enter the input–output flow map. We can now proceed to define formally the problem.

Problem 1 Let Hypothesis 2 hold, and fix $f^* \in \mathbf{BV}([0, T], [f^{min}, f^{max}])$ and $K > 0$. Find the control policy $v \in \mathbf{BV}([0, T], [v_{min}, v_{max}])$, with $\mathrm{TV}(v) \leq K$, which minimizes the functional $J : \mathbf{BV}([0, T], [v_{min}, v_{max}]) \to \mathbb{R}$ defined by

$$J(v) := \int_0^T (Out(t) - f^*(t))^2 dt \tag{4.13}$$

where $Out(t)$ is given by (4.12).

Remark 9 We point out that even if the problem is addressed in free flow condition, it is not possible to find explicit solutions to the problem. However, if In, Out, f^*, and u_0 are constant in time, i.e., do not depend on time, the problem has a trivial solution which is $v = \dfrac{f^*}{u_0}$ that gives $J(v) = 0$.

4.3.2   Needle-Like Variations

In this section, we estimate the variation of the cost $J(v)$ with respect to the perturbations of the control policy v. In this way, we can prove continuous dependence of the solution from the control policy.

Let us fix the notation for integrals of BV function with respect to Radon measures.

Definition 16 Let ϕ be a BV function and μ a Radon measure. We define

$$\int \phi(x^+)\, d\mu(x) := \int \phi(x)\, d\mu_c(x) + \sum_i m_i \phi(x_i^+),$$

where $\mu = \mu_c + \sum_i m_i \delta_{x_i}$ is the decomposition of μ into its continuous and Dirac parts.

Remark 10 We recall that any Radon measure on $\mathbb{R}$ can be decomposed into its continuous (AC+Cantor) and Dirac parts, as a consequence of the Lebesgue decomposition theorem; for more details, see, e.g., [127].

We produce a variation of the value of $v(\cdot)$ on small intervals of the type $[t, t + \Delta t]$ in the same spirit as the needle variations of Pontryagin Maximum Principle [55] and compute the variation in the cost. The analytical expression of variations will allow to compute analytically a gradient and hence to implement a steepest descent type strategy to find the optimal speed limit.

Definition 17 Consider $v \in \mathbf{BV}([0, T], [v_{\min}, v_{\max}])$, with T sufficiently large so that $t_0 < \tau(T)$, and a time t such that $\tau^{-1}(0) = t_0 \leq t < \tau(T)$ and $v(t^+) < v_{\max}$. Let $\Delta v > 0$ and $\Delta t > 0$ be sufficiently small such that $t + \Delta t \leq \tau(T)$ and $v(t^+) + \Delta v \leq v_{\max}$. We define a needle-like variation $v'(\cdot)$ of v, corresponding to t, Δt, and Δv by setting $v'(s) = v(s) + \Delta v$ if $s \in [t, t + \Delta t]$ and $v'(s) = v(s)$ otherwise, see Fig. 4.3.

Lemma 4 *Consider $v \in \mathbf{BV}([0, T], [v_{\min}, v_{\max}])$, and let v' be a needle-like variation of v. Then, it holds*

Fig. 4.3 Needle-like variation of the velocity v

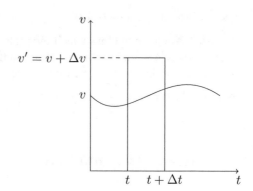

$$\lim_{\Delta v \to 0^+} \lim_{\Delta t \to 0^+} \frac{J(v') - J(v)}{\Delta v} =$$

$$= 2u^2(t, L^-)v(t^+) - 2u(t, L^-)f^*(t^+) +$$

$$- \int_{]0,L]} v((t + s(x))^+)\, du_x^2(t) + 2 \int_{]0,L]} f^*((t + s(x))^+))\, du_x(t) +$$

$$+ 2\frac{\ln(t^-)}{v(t^+)} \left(f^*(t^+) - \frac{v(\tau^{-1}(t')^-)}{v(t^+)} \ln(t^-) \right),$$

$$(4.14)$$

*where integrals are defined according to (16). For $\Delta v < 0$, the limit for $\Delta v \to 0^-$
satisfies the same formula with right limits replaced by left limits in the two integral
terms in (4.14).*

For the proof, the reader is referred to [116].
The results, shown in Lemma 4, allow us to prove the following existence result.

Proposition 10 *Problem* (1) *admits a solution.*

Proof The space $\Omega = \{v \in \mathbf{BV}([0, T], [v_{\min}, v_{\max}]) : TV(v) \leq K\} \cap \{v \in L^\infty([0, T], [v_{\min}, v_{\max}]) : \|v\|_\infty \leq C\}$ is compact in $\mathbf{L}^1$, see, e.g., [8], and J is
Lipschitz continuous on Ω, and thus there exists a minimizer of (1). □

4.3.3 Three Different Control Policies

In this section, we define three control policies for the time-dependent maximal
speed v. The first, called the instantaneous policy (IP), is defined by minimizing
the instantaneous contribution for the cost $J(v)$ at each time. Then, we introduce
a second control policy, called random exploration (RE) policy. Such policy uses a
random path along a binary tree, which corresponds to the upper and lower bounds
for v, i.e., $v = v_{\max}$ and $v = v_{\min}$.
Finally, a third control policy is searched using a gradient descent method (GDM).
Compared with classical GDM methods, in this section, a different approach is used
and the gradient is replaced with cost variations computed with respect to needle-
like variations in the control policy. The key ingredient to define the third policy is
the explicit computation of the gradient that was computed in [116], while the other
two policies are chosen such that they will provide a fair comparison with respect
to the state of the art.

4.3.3.1 Instantaneous Policy

Definition 18 Consider (1). Define the **instantaneous policy** as follows:

$$v(t) := P_{[v_{\min}, v_{\max}]}\left(f^*(t^-) \cdot \frac{v(\tau(t)^-)}{\ln(\tau(t)^-)} \right), \tag{4.15}$$

where the projection $P_{[v_{\min}, v_{\max}]} : \mathbb{R} \to \mathbb{R}$ is the function

$$P_{[a,b]}(x) := \begin{cases} a & \text{for } x < a, \\ x & \text{for } x \in [a, b], \\ b & \text{for } x > b. \end{cases} \tag{4.16}$$

4.3.3.2   Random Exploration Policy

The random exploration policy is defined as follows:

Definition 19 Given the extreme values for the maximal speed, $v_{\max}$ and $v_{\min}$, and a time step Δt, the **random exploration policy** draws sequences of velocities from the set $\{v_{\max}, v_{\min}\}$ corresponding to control policy values for each Δt.

4.3.3.3   Gradient Method

We use needle-like variations and the analytical expression in (4.14) to numerically compute one-sided variations of the cost. We consider such variations as estimates of the gradient of the cost in L^1. More precisely, we give the following definition.

Definition 20 The **gradient policy** is the result of a first order optimization algorithm to find a local minimum to (1) using the Gradient Descent Method and the expression in (4.14), stopping at a fixed precision tolerance.

4.3.4   Numerical Results

In this section, we show the numerical results obtained by implementing the three different policies. The numerical algorithm for all the approaches is composed of two steps:

1. Numerical scheme for the conservation law. The density values are computed using the classical Godunov scheme, introduced in [154].
2. Numerical solution for the optimal control problem, i.e., computation of the maximal speed using the instantaneous control, random exploration policy, and gradient descent.

Let Δx and Δt be the fixed space and time steps, and set $x_{j+\frac{1}{2}} = j \Delta x$, the cell interfaces such that the computational cell is given by $C_j = [x_{j-\frac{1}{2}}, x_{j+\frac{1}{2}}]$. The center of the cell is denoted by $x_j = (j - \frac{1}{2})\Delta x$ for $j \in \mathbb{Z}$ at each time step $t^n = n\Delta t$ for $n \in \mathbb{N}$. We fix $\mathcal{J}$ the number of space points and T the finite time horizon. We now describe in detail the two steps.

4.3.4.1 Godunov Scheme for Hyperbolic PDEs

The Godunov scheme can be expressed in conservative form as

$$u_j^{n+1} = u_j^n - \frac{\Delta t}{\Delta x}\left(F(u_j^n, u_{j+1}^n, v^n) - F(u_{j-1}^n, u_j^n, v^n)\right), \tag{4.17}$$

where v^n is the maximal speed at time t^n. Additionally, $F(u_j^n, u_{j+1}^n, v^n)$ is the Godunov numerical flux that in general has the following expression:

$$F(u_j^n, u_{j+1}^n, v^n) = \begin{cases} \min_{z \in [u_j^n, u_{j+1}^n]} f(z, v^n) & \text{if } u_j^n \leq u_{j+1}^n, \\ \max_{z \in u_{j+1}^n, u_j^n} f(z, v^n) & \text{if } u_{j+1}^n \leq u_j^n. \end{cases} \tag{4.18}$$

For clarity, the maximal velocity was included as an argument for the Godunov scheme so that the dependence of the scheme on the optimal control could be explicit.

4.3.4.2 Velocity Policies

The next step in the algorithm consists of computing a control policy v that can be used in the Godunov scheme with the different policies.

- **Instantaneous policy**
  At each time step, the velocity v^{n+1} is computed using the following formula:

$$v^{n+1} = v(t^{n+1}) = P_{[v_{\min}, v_{\max}]}\left(\frac{f^*(t^n)}{u_{\mathcal{J}}^n}\right). \tag{4.19}$$

- **Random exploration policy**
  To compute for each time step the value of the velocity, a randomized path on a binary tree, see Fig. 4.4, is used. With such technique, we obtain several sequences of possible velocities. For each sequence, the velocities are used to compute the fluxes for the numerical simulations. We then choose the sequence that minimizes the cost.

  Remark 11 Notice that the control policy RE may have a very large total variation, and thus it might not respect the bounds on TV given in (1). Therefore, the found control policies may not be allowed as a solution of this problem.

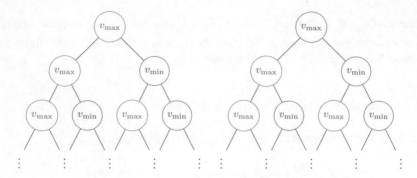

Fig. 4.4 The first branches of the binary tree used for sampling the velocity

However, this technique was implemented for comparison with the results and performances obtained by the GDM.

- **Gradient descent method** First, numerically one-sided variations of the cost are computed using (4.14). Then, the classical gradient descent method [4] is used to find the optimal control strategy and to compute the optimal velocity that fits the given outflow profile.

4.3.4.3   Simulations

The following parameters: $L = 1$, $\mathcal{J} = 100$, $T = 15.0$, $u_{cr} = 0.5$, $u_{max} = 1$, $v_{min} = 0.5$, and $v_{max} = 1.0$ are chosen. Moreover, the input flux at the boundary of the domain is given by In $= \min(0.3 + 0.3\sin(2\pi t^n), 0.5)$. Two different target fluxes $f^* = 0.3$ and $f^* = |(0.4\sin(t\pi - 0.3))|$ are used. The initial condition is a constant density $u(0, x) = 0.4$, and oscillating inflows to represent variations in typical inflow of urban or highway networks at the 24 h time scale are given.

Test I: Constant Outflow

In Fig. 4.5, the time-varying speed obtained by using the three policies is shown. In each case, we notice that due to the oscillating input signal the control policy is also oscillating. From a practical point of view, a solution where the speed changes at each time step might be unfeasible, but these policies could be seen as periodic change of maximal speed for different time frames during the day when the time horizon is scaled to the day length. In Table 4.1, one can see the different results obtained for the cost functional computed at the final time for the different policies. The instantaneous policy is outperformed by the random exploration policy and by the gradient method. In Fig. 4.6, one can see the distribution of the different values of the cost functional over 1000 simulations. Moreover, in Fig. 4.7, one can see the

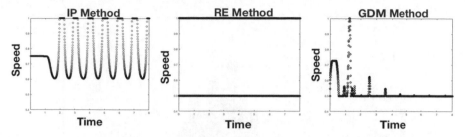

Fig. 4.5 Speed obtained by using the instantaneous policy (left), random exploration policy (center), and the gradient descent method (right) for a target flux $f^* = 0.3$

Table 4.1 Value of the cost functional and the average velocity for the different policies

Method	Cost functional	Average speed
Fixed speed $v = v_{max} = 1.0$	873.0786	1.0
Fixed speed $v = v_{min} = 0.5$	785.2736	0.5
Instantaneous policy	850.3704	0.7867
Minimum of random exploration policy	723.6733	0.7597
Gradient method	735.0565	0.5241

differences between the actual outflow obtained and the target one for all methods. The CPU time for the different simulations approaches (see Table 4.2) is compared, and as expected, the random exploration policy is the least performing, while the instantaneous policy is the fastest one. In addition, one can look at the TV(v) for each one of the policies obtaining the following results:

- IP: TV(v) = 12.6904
- RE: TV(v) = 753.5
- GDM: TV(v) = 70.81333

Notice that the simple case of a fixed speed has TV(v) = 0, making this option the most performing from this point of view.

Test II: Sinusoidal Outflow

In Fig. 4.8, it is shown the optimal velocity obtained by using the instantaneous policy, the random exploration, and the gradient descent method with a sinusoidal outflow. One can see that the different policies give different profiles of optimal speed. In each case, we can see that an a posteriori treatment of the speeds before implementation in real traffic might be needed. Figure 4.9 shows the histogram of the cost functional obtained for the random exploration policy, and in Fig. 4.10, the real outgoing flux with the target one is compared. In Table 4.3, different results obtained for the cost functional computed at final time for the different policies are shown. Also, in this case, the instantaneous policy is outperformed by the other two. The CPU times give results similar to the previous test.

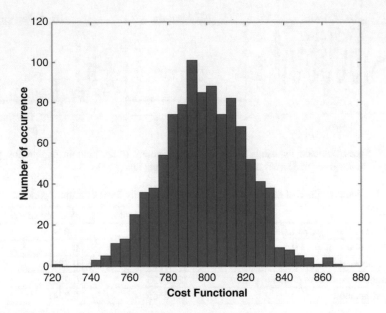

Fig. 4.6 Histogram of the distribution of the value of the cost functional for the random exploration policy. We run 1000 different simulations

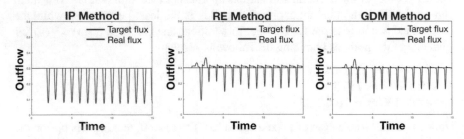

Fig. 4.7 Difference between the real outgoing flux and the target constant flux, computed with the instantaneous policy (left), the gradient method (right) and the random exploration policy (center)

Table 4.2 CPU time for the simulations performed with the different approaches

Method	CPU time (s)
Instantaneous policy	32.756
Random exploration policy	7577.390
Gradient method	1034.567

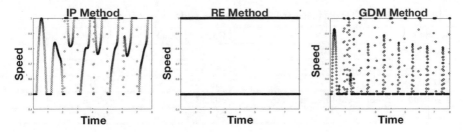

Fig. 4.8 Speed obtained by using the instantaneous policy (left), the gradient descent method (right), and the random exploration policy (center) for a sinusoidal target flux

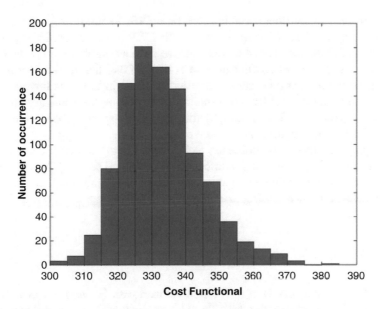

Fig. 4.9 Histogram of the distribution of the value of the cost functional for the random exploration policy. We run 1000 different simulations

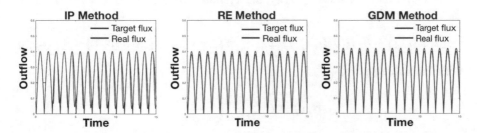

Fig. 4.10 Difference between the real outgoing flux and the target sinusoidal flux, computed with the instantaneous policy (top left), the gradient method (top right), and the random exploration policy (bottom)

Table 4.3 Value of the cost functional for the different policies

Method	Cost functional	Average speed
Fixed speed $v = v_{max} = 1.0$	$1.3979e + 03$	1.0
Fixed speed $v = v_{min} = 0.5$	843.3395	0.5
Instantaneous policy	458.8874	0.7917
Minimum of random exploration policy	303.8327	0.7512
Gradient method	307.6889	0.6001

4.4   Discrete-Optimization Methods for First Order Models

In this section, we consider variable speed limit (VSL) control problem coupled to ramp metering for first order equations, i.e., the LWR model. In particular, merging these two problems means that on one side we have a conservation law with time-dependent discontinuous coefficients due to the fact that the maximal velocity for the VSL problem must be evaluated at discrete points in time. On the other hand, the on-ramp metering problem corresponds to controlling the boundary conditions at junctions [143, 144]. The resulting optimal control problem is then a combination of two controls directly influencing each other.

We apply continuous optimization techniques, where the first order optimality system is derived and solved by a descent type method [188, 257]. More precisely, we work on the approximate dynamics, thus resulting in the so-called *discretize-then-optimize approach* leading to a finite-dimensional optimality system [160].

4.4.1   Traffic Flow Network Modeling

As detailed in Appendix B, a traffic flow network can be modeled as a directed graph $\mathcal{G} = (\mathcal{I}, \mathcal{J})$, where the edges $\mathcal{I} = \{I_\ell\}_\ell$ correspond to roads and the vertices $\mathcal{J} = \{J_j\}_j$ to junctions or intersections. Each edge $I_\ell \in \mathcal{I}$ is represented by an interval $[0, L_\ell]$ and $u_\ell(x, t)$ denotes the density of cars on road I_ℓ.

Given initial conditions $u_\ell(x, 0)$, the dynamics on the network is described by the LWR equation

$$\partial_t u_\ell(t, x) + \partial_x f_\ell(u_\ell(t, x), t) = 0 \qquad \forall \ell, \ x \in]0, L_\ell[, \ t \in [0, T] \qquad (4.20)$$

with Greenshields flux

$$f_\ell(t, u) = u \, v^\ell_{max}(t) \left(1 - \frac{u}{u^\ell_{max}}\right),$$

where $v^\ell_{max}(t)$ is the (piecewise constant) maximal speed limit and u^ℓ_{max} is the maximal car density corresponding to the jammed situation, see, for instance, Fig. 4.11. The maximal flux $f^{max}_\ell(t)$ is attained at $u = u^\ell_{cr} = u^\ell_{max}/2$, see Fig. 4.11.

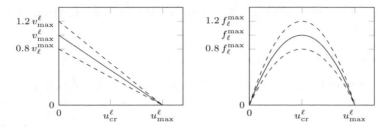

Fig. 4.11   Velocity and flow rate for different values of $v^\ell_{\max}(t)$

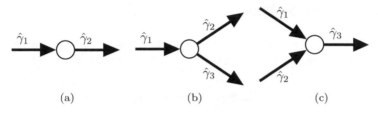

(a)   (b)   (c)

Fig. 4.12   Different types of junctions. (**a**) One-to-one. (**b**) Diverging. (**c**) Merging

4.4.1.1   Coupling Conditions at Junctions

The dynamics at junction nodes $J_j \in \mathcal{J}$ must guarantee mass conservation and is defined using the demand and supply functions corresponding to $f_\ell(t, u)$:

$$
D_\ell(u, t) = \begin{cases} f_\ell(u, t) & \text{if } u \leq u^\ell_{\mathrm{cr}}, \\ f^{\max}_\ell(t) & \text{if } u \geq u^\ell_{\mathrm{cr}}, \end{cases} \quad \text{and} \quad S_e(u, t) = \begin{cases} f^{\max}_\ell(t) & \text{if } u \leq u^\ell_{\mathrm{cr}}, \\ f_\ell(u, t) & \text{if } u \geq u^\ell_{\mathrm{cr}}. \end{cases}
$$

In the rest of this section, only the cases of one-to-one, merging, and diverging junctions are considered, i.e., Fig. 4.12.

The coupling conditions are computed as follows:

- One-to-one junction: the fluxes at the junction are obtained by

$$
\hat\gamma_1 = \hat\gamma_2 = \min\{D_1(u_1, t), S_2(u_2, t)\} . \tag{4.21}
$$

- Diverging junction: the distribution of cars on outgoing roads is described by the parameters $\alpha_{2,1} \geq 0$ and $\alpha_{3,1} \geq 0$ such that $\alpha_{2,1} + \alpha_{3,1} = 1$, and the fluxes at the junction are given by the following (non-FIFO) conditions [176, 205]:

$$
\hat\gamma_2 = \min\{\alpha_{2,1}D_1(u_1, t), S_2(u_2, t)\} , \tag{4.22a}
$$

$$
\hat\gamma_3 = \min\{\alpha_{3,1}D_1(u_1, t), S_3(u_3, t)\} , \tag{4.22b}
$$

$$
\hat\gamma_1 = \hat\gamma_2 + \hat\gamma_3 . \tag{4.22c}
$$

- Merging junction: we introduce a priority parameter $P \in (0, 1)$, and the fluxes are given by

$$\hat{\gamma}_1 = \min \{D_1(u_1, t), \max \{PS_3(u_3, t), S_3(u_3, t) - D_2(u_2, t)\}\} , \qquad (4.23a)$$

$$\hat{\gamma}_2 = \min \{D_2(u_2, t), \max \{(1 - P)S_3(u_3, t), S_3(u_3, t) - D_1(u_1, t)\}\} , \qquad (4.23b)$$

$$\hat{\gamma}_3 = \hat{\gamma}_1 + \hat{\gamma}_2 . \qquad (4.23c)$$

4.4.1.2 Boundary Conditions

If an arc I_ℓ that is connected to the network only downstream while the desired inflow rate upstream is $f_\ell^{\text{in}}(t)$, the actual inflow to the road I_ℓ is given by

$$\gamma_\ell^{\text{in}} = \min \left\{ f_\ell^{\text{in}}(t), S_\ell(u_\ell, t) \right\} . \qquad (4.24)$$

Assuming instead that there is a queue at the upstream node of arc I_ℓ with length $l_\ell(t)$, then, the inflow to the road is given by

$$\gamma_\ell^{\text{in}} = \min \{D_\ell(l_\ell, t), S_\ell(u_\ell, t)\} , \qquad (4.25)$$

where the demand function depends on the length $l_\ell(t)$ of the queue at time t,

$$D_\ell(l_\ell, t) = \begin{cases} \tilde{f}_\ell^{\text{max}} & \text{if } l_\ell > 0, \\ \min \left\{ f_\ell^{\text{in}}(t), f_\ell^{\text{max}} \right\} & \text{if } l_\ell = 0. \end{cases} \qquad (4.26)$$

Above, $\tilde{f}_\ell^{\text{max}}$ denotes the maximum flux that can enter the road from the queue. The evolution of the queue length $l_\ell(t)$ is given the ODE

$$\frac{dl_\ell(t)}{dt} = f_\ell^{\text{in}}(t) - \gamma_\ell^{\text{in}} \qquad (4.27)$$

for an initial state $l_\ell(0)$.

Conversely, on arcs connected only upstream to the network, absorbing boundary conditions are prescribed up to a given maximum flow rate $f_\ell^{\text{out}}(t)$ as

$$\gamma_\ell^{\text{out}} = \min \left\{ f_\ell^{\text{out}}(t), D_\ell(u_\ell, t) \right\} , \qquad (4.28)$$

thus ensuring that cars exit the network if their flow is lower than the maximum flow rate.

4.4.2 Optimization Problem for VSL and Ramp Metering

In this section, we describe the optimization problem, consisting of minimizing the total travel time [267] and/or maximizing the outflow of the system, controlling traffic flow through a network by adjusting maximal speed limits and on-ramp fluxes:

$$
\min J(\vec{l}, \vec{u}, \vec{\gamma}^{\text{out}}) := \sum_\ell \beta_\ell \int_0^T \left(l_\ell(t) + \int_0^{L_\ell} u_\ell(x, t)\, dx \right) dt - \sum_\ell \varepsilon_\ell \int_0^T \gamma_\ell^{\text{out}}(t)\, dt,
$$

(4.29)

where we set $\vec{l} = (l_\ell)_\ell$, $\vec{u} = (u_\ell)_\ell$ and $\vec{\gamma}^{\text{out}} = (\gamma_\ell^{\text{out}})_\ell$. Alternatively, one can choose to minimize a congestion measure as in [247], i.e.,

$$
\min J(\vec{l}, \vec{u}) := \sum_\ell \beta_\ell \int_0^T \left(l_\ell(t) + \max \left\{ 0, \int_0^{L_\ell} \left(u_\ell(x, t) - \frac{f_\ell(u_\ell(x, t), t)}{v_{\ell, \text{ref}}} \right) dx \right\} \right) dt.
$$

(4.30)

Above, β_ℓ and ε_ℓ denote non-negative weights and $v_{\ell,\text{ref}}$ a reference velocity.

4.4.2.1 Variable Speed Limits

We assume that the time-dependent maximal velocities $v_{\text{max}}^\ell(t)$ have uniform lower and upper bounds

$$
v_{\text{low}}^\ell \leq v_{\text{max}}^\ell(t) \leq v_{\text{high}}^\ell \qquad \forall t \in [0, T].
$$

To get a finite number of speed changes, we introduce control points $v^k \in [0, T]$, $k \in \{0, \ldots, Nu\}$, and the corresponding control variables V_ℓ^k for each road. The maximal speed $v_{\text{max}}^\ell(t)$ is assumed to be piecewise constant on the control grid:

$$
v_{\text{max}}^\ell(t) = V_\ell^{k+1} \qquad \forall t \in \,]v^k, v^{k+1}].
$$

(4.31)

4.4.2.2 Ramp Metering

On-ramp dynamics can be described in the framework of merging junctions, where we set the index $\ell = 2$ for the on-ramp, which will be described as a queue (4.26). Here, the aim is to control the main lane access from the on-ramp through a controlled demand function that is defined as

$$D_\ell^c(l_\ell, t) = \omega_\ell(t) D_\ell(l_\ell, t), \tag{4.32}$$

where $D_\ell(l_\ell, t)$ is given by (4.26). Then, plugging (4.32) in (4.23), we obtain

$$\hat{\gamma}_1 = \min\left\{ D_1(u_1, t), \max\left\{ P S_3(u_3, t), S_3(u_3, t) - D_2^c(l_2, t) \right\} \right\}, \tag{4.33a}$$

$$\hat{\gamma}_2 = \min\left\{ D_2^c(l_2, t), \max\left\{ (1 - P) S_3(u_3, t), S_3(u_3, t) - D_1(u_1, t) \right\} \right\}, \tag{4.33b}$$

$$\hat{\gamma}_3 = \hat{\gamma}_1 + \hat{\gamma}_2. \tag{4.33c}$$

Correspondingly, the evolution of the on-ramp buffer changes to

$$\frac{dl_2(t)}{dt} = f_2^{\text{in}}(t) - \hat{\gamma}_2, \tag{4.34}$$

where $f_2^{\text{in}}(t)$ is the external boundary inflow at the on-ramp.

For ramp metering, we consider a piecewise constant control function:

$$\omega_\ell(t) = \omega_\ell^{k+1} \qquad \forall t \in (\nu^k, \nu^{k+1}]. \tag{4.35}$$

Finally, the joint speed limit and ramp metering control problem for traffic flow on networks is given by

$$\min_{\vec{z}, \vec{\omega}} J(\vec{l}, \vec{u}, \vec{\gamma}^{\text{out}}) \tag{4.36}$$

$$\text{s.t. } (4.20)\text{--}(4.28), (4.31), (4.32)\text{--}(4.35)$$

with $\vec{z} = (z_\ell)_\ell$ and $\vec{\omega} = (\omega_\ell)_\ell$.

4.4.3   Numerical Simulations

Consider a time mesh $t^n = n \Delta t$ with $\Delta t = \frac{T}{Nt}$ and divide each road ℓ into Nx_ℓ cells of size $\Delta x_\ell = \frac{L_\ell}{Nx_\ell}$.

The discretized objective functions corresponding to (4.29) and (4.30) are

$$\min \sum_\ell \beta_\ell \sum_{n=1}^{Nt} \left(l_\ell^n + \sum_{j=1}^{Nx_\ell} u_{\ell, j-0.5}^n \Delta x_\ell \right) \Delta t - \sum_\ell \varepsilon_\ell \sum_{n=1}^{Nt} f(u_{\ell, Nx_\ell}^n, t^n) \Delta t, \tag{4.37}$$

and

$$\min \sum_{\ell} \beta_\ell \sum_{n=1}^{Nt} \left(l_\ell^n + \max \left\{ 0, \sum_{j=1}^{Nx_\ell} \left(u_{\ell,j-0.5}^n - \frac{f_\ell(u_{\ell,j-0.5}^n, t^n)}{v_{\ell,\mathrm{ref}}} \right) \Delta x_\ell \right\} \right) \Delta t ,$$

(4.38)

where $v_{\ell,\mathrm{ref}} = \frac{1}{2} v_\ell^{\mathrm{high}}$. The conservation law (4.20) is discretized using a staggered Lax–Friedrichs scheme [216]

$$u_{0.5}^{n+1} = \tfrac{1}{4} \left(3u_{0.5}^n + u_{1.5}^n \right) - \tfrac{\lambda}{2} \left[f(u_{1.5}^n, t^n) + f(u_{0.5}^n, t^n) - 2f(u_0^n, t^n) \right] ,$$

(4.39a)

$$u_{j-0.5}^{n+1} = \tfrac{1}{4}(u_{j-1.5}^n + 2u_{j-0.5}^n + u_{j+0.5}^n) - \tfrac{\lambda}{2}[f(u_{j+0.5}^n, t^n) - f(u_{j-1.5}^n, t^n)] ,$$

(4.39b)

$$u_{Nx-0.5}^{n+1} = \tfrac{1}{4}(u_{Nx-1.5}^n + 3u_{Nx-0.5}^n)$$
$$- \tfrac{\lambda}{2}[2f(u_{Nx}^n, t^n) - f(u_{Nx-0.5}^n, t^n) - f(u_{Nx-1.5}^n, t^n)] ,$$

(4.39c)

where $\lambda = \Delta t / \Delta x$ (skipping the index ℓ) and

$$u_{\ell,j-0.5}^n \approx \frac{1}{\Delta x_\ell} \int_{(j-1)\Delta x_\ell}^{j\Delta x_\ell} u_\ell(x, t^n)\, dx \qquad \text{for} \quad j \in \{1, \dots, Nx_\ell\},\ n \in \{0, \dots, Nt\} .$$

4.4.3.1 Optimization Approach

The discrete (finite-dimensional) optimization problem (4.37) is solved with an SQP solver (DONLP2) [258, 259], requiring gradient computation. This is achieved using the adjoint approach, which is recalled below.

Given an objective function $J(W, Y)$ (here (4.37)), where W denotes the control variables of the discretized model equations (the speed limits z_ℓ^k and the ramp controls ω_ℓ^k) and Y the state variables (densities u_ℓ^n, flow rates $f(u_\ell^n)$, queue lengths l_ℓ^n). The discretized model equations are denoted $E(W, Y) = 0$. Assuming that the model equations $E(W, Y) = 0$ have a unique solution $Y = Y(W)$ for any fixed W, we denote by $J(W) = J(W, Y(W))$ the reduced problem. Let ξ be the solution of the adjoint equation

$$\left(\frac{\partial}{\partial Y} E(W, Y(W)) \right)^T \xi = - \left(\frac{\partial}{\partial Y} J(W, Y(W)) \right)^T ,$$

(4.40)

and the cost gradient can be efficiently computed as

$$\frac{d}{du} J(W, Y(W)) = \frac{\partial}{\partial W} J(W, Y(W)) + \xi^T \frac{\partial}{\partial W} E(W, Y(W)) .$$

(4.41)

4.4.3.2   Numerical Results

This section collects the numerical results corresponding to an example of variable speed limit control and a combined optimization of variable speed limits (VSLs) and ramp metering.

Variable Speed Limit (VSL) Control

The topology of the considered road network (including distribution rates α and priority parameters P) is shown in Fig. 4.13. Further road properties as well as the initial conditions are given in Table 4.4. Additionally, Fig. 4.14 shows the inflow profiles within the considered time horizon of 1000 seconds.

   For the described setting, we fix the discretization parameters and an increasing number of control points Nu. We separately minimize the total travel time (with $\beta_\ell = 10^{-3}$) and maximize the accumulated outflow at the nodes outA and outB (with $\varepsilon_\ell = 10^{-1}$). The results are collected in Table 4.5 and Fig. 4.15, showing that lower travel times/larger outflows are obtained for an increasing number of control points. In particular, compared to the uncontrolled case, where all speed limits are taken at the upper bound, there is an improvement of 1.28%/0.03%.

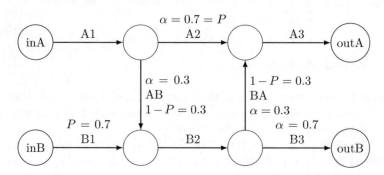

Fig. 4.13   Road network with two main roads

Table 4.4   Properties of the roads in Fig. 4.13

Road	Length	u_{max}	v_{low}	v_{high}	Initial density
A1	1000	2	20	30	0.3
A2	1000	2	20	30	0.3
A3	1000	2	20	30	0.3
B1	1000	2	20	30	0.3
B2	1000	2	20	30	1.2
B3	1000	2	20	30	1.2
AB	200	0.5	10	20	0.1
BA	200	0.5	10	20	0.1

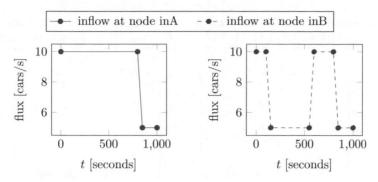

Fig. 4.14 Inflow profiles for the network in Fig. 4.13

Table 4.5 Optimal travel times/outflows for different numbers of control points

Nu	$\Delta x / \Delta t$	Min. travel time	Max. outflow
–	50 / 0.5	2336.698921	1844.762416
5	50 / 0.5	2307.370375	1845.023309
20	50 / 0.5	2306.863796	1845.289234
80	50 / 0.5	2306.694086	1845.353672

We then compare the results with a fixed discretization but an additional penalty term with weight $\delta_\ell \geq 0$ in (4.37):

$$\sum_e \delta_e \sum_{n=1}^{Nt} \Delta t \left(\frac{v_e^{\max}(t^n) - v_e^{\max}(t^{n-1})}{v_e^{\text{high}} \Delta t} \right)^2 ,$$

which penalizes variations in the control. The resulting optimal controls are provided in Fig. 4.16. Obviously, the additional penalty term leads to smoother optimal solutions.

VSL and Coordinated Ramp Metering

In this section, we combine the optimization of variable speed limits and ramp metering. The considered network is shown in Fig. 4.17, and the corresponding parameters are given in Table 4.6. The priority parameter at the on-ramp is $P = 0.5$ and we take $\tilde{f}^{\max} = 1.5$. Figure 4.18 shows the inflow profiles and the maximum outflow. We take $Nu = 36$ control points and the cost functional (4.38), with the following constraints: a maximum queue size of 50 cars at the entrance of the main road (node "in") and a maximum queue size of 600 cars at the on-ramp are allowed.

In the uncontrolled case (Fig. 4.19), the inflow to the main road from the on-ramp varies from full inflow at the beginning of the simulation ($f^{\text{in}} = 0.75$) to $P \cdot f^{\text{out}} = 0.5$ due to the congestion at road 2 (outflow at the node "out" is 1.0 and priority parameter $P = 0.5$). The flow on the on-ramp then increases again up to the

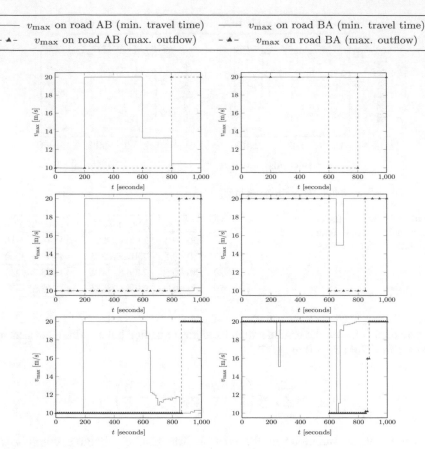

Fig. 4.15 Optimal control of v_{max} for an increasing number of control points

maximum flow level ($\tilde{f}^{max} = 1.5$) when the maximum possible outflow at the node "out" increases. Finally, the inflow to the main road from the on-ramp decreases to the inflow at this node again ($f^{in} = 0.75$) when the queue at the on-ramp is empty.

Figures 4.20, 4.21, and 4.22 show the computed optimal controls, the queue lengths at the inflow of the main road and at the on-ramp in the uncontrolled and the optimized cases, and the density at the beginning of road 2, respectively. The speed control is activated during large inflow periods, while the on-ramp control acts on the queue length. Moreover, in the optimal solution, the queue at the entrance of the main road is kept empty, while the queue at the on-ramp displays a large peak when traffic gets congested after about 2 hours due to the reduced density on the main road.

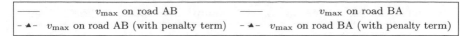

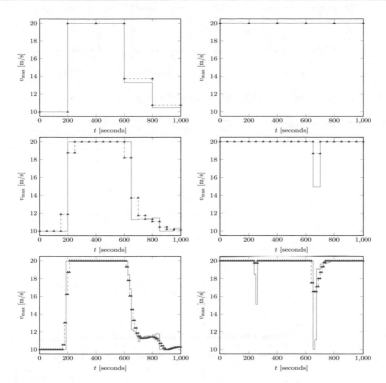

Fig. 4.16 Optimal control of v_{max} with and without penalty term (min. travel time)

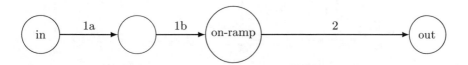

Fig. 4.17 Road network with an on-ramp at the node "on-ramp"

Table 4.6 Properties of the roads in Fig. 4.17

Road	Length	u_{max}	v_{low}	v_{high}	Initial density
1a	2000	2	30	30	0.1
1b	2000	2	10	30	0.1
2	4000	2	30	30	0.1

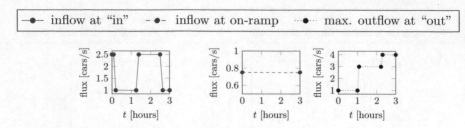

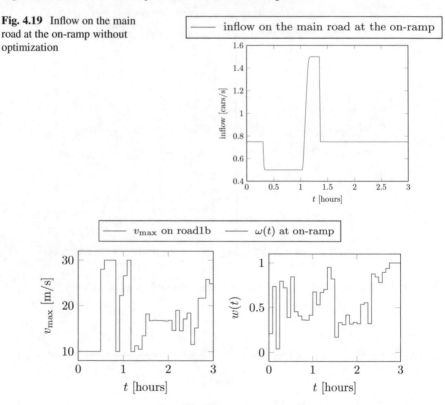

Fig. 4.18 Inflow/max. outflow profiles for the network in Fig. 4.17

Fig. 4.19 Inflow on the main road at the on-ramp without optimization

Fig. 4.20 Optimal control of $v_{max}(t)$ on road 1b and $\omega(t)$ at the on-ramp

4.5 Discrete-Optimization Methods for Second Order Models

In this section, we focus on coupling conditions of road networks at on-ramps, where capacity drops usually occur for high density traffic, see, e.g., [169, 183, 211, 239, 260]. By capacity drop we indicate that the measured outflow of the system is smaller than what it could be in optimal conditions, with differences up to 10% or more [183, 260]. This is due to inefficient driving reaction at the exit of a congested

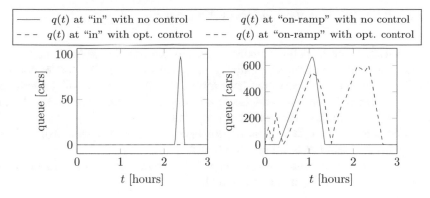

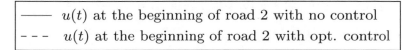

Fig. 4.21 Queue at the node "in" and the on-ramp

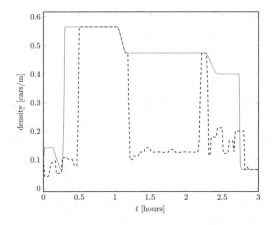

Fig. 4.22 Density at the beginning of road 2 with and without optimization

zone (upstream the on-ramp), which reduces the downstream flow compared to free flow conditions [267].

While this cannot be captured by first order models like LWR, the challenge for second order models is to find appropriate conditions which capture the capacity drop phenomenon, while ensuring the conservation of mass and momentum flow. Once these coupling conditions are defined, they can be integrated in a finite-volume type numerical scheme to compute the evolution of traffic conditions on the network.

4.5.1 The Aw–Rascle Model on Networks

Let us introduce the modeling equations given by the Aw–Rascle (AR) model [23] including a relaxation term as originally proposed in [157]. The model consists of a 2×2 system of conservation laws for the density and a sort of *generalized momentum* derived from the (anticipate) acceleration equation. We explain below how it can be extended to the context of networks.

Traffic states are described by the density $u_i(x, t)$ and the mean speed of vehicles $v_i(x, t)$ on each road i at position x and time t.

Given some initial state $(u_i(x, 0), v_i(x, 0))$ on each road i, the dynamics for $x \in \,]0, L_i[$ and $t \in \,]0, T[$ are described by [23, 157]

$$
\begin{cases}
\partial_t u_i + \partial_x(u_i v_i) = 0, \\
\partial_t(u_i(v_i + p_i(u_i))) + \partial_x(u_i v_i(v_i + p_i(u_i))) = -u_i \frac{v_i - V_i(u_i)}{\delta},
\end{cases}
\tag{4.42}
$$

or, setting $y_i = u_i(v_i + p_i(u_i))$,

$$
\partial_t \begin{pmatrix} u_i \\ y_i \end{pmatrix} + \partial_x \begin{pmatrix} y_i - u_i p_i(u_i) \\ (y_i - u_i p_i(u_i)) \frac{y_i}{u_i} \end{pmatrix} = \underbrace{\begin{pmatrix} 0 \\ -\frac{(y_i - u_i p_i(u_i)) - u_i V_i(u_i)}{\delta} \end{pmatrix}}_{= g_i(u_i, y_i)},
\tag{4.43}
$$

where $p_i(u)$ is a given pressure function satisfying $p_i'(u) > 0$ and $u p_i''(u) + 2 p_i'(u) > 0$ for all u. The latter condition ensures that, for any constant $c > 0$, the curve $\{v + p_i(u) = c\}$ is strictly concave and passes through the origin of the (u, uv)-plane. Moreover, there exists a unique point $\sigma_i(c)$ maximizing the flux uv along the curve $\{v + p_i(u) = c\}$. The relaxation term in the momentum equation expresses the tendency of drivers to adjust their velocity to a desired speed $V_i(u)$, with a relaxation time $\delta > 0$.

Aiming at traffic control applications, in the following, we consider the time-dependent preferential velocity:

$$
V_i(u, t) = v_{\max}^i(t) \left(1 - \frac{u}{u_{\max}^i} \right)
\tag{4.44}
$$

and the pressure function (as in [169, 239])

$$
p_i(u, t) = \frac{v_i^{\text{ref}}(t)}{\gamma_i} \left(\frac{u}{u_{\max}^i} \right)^{\gamma_i}
\tag{4.45}
$$

equipped with maximal density $u_{\max}^i > 0$, maximal velocity $v_{\max}^i(t) > 0$, reference velocity $v_i^{\text{ref}}(t) > 0$, and $\gamma_i > 0$. Therefore, also the source term in (4.43) becomes time-dependent: $g_i(u, y) = g_i(u, y, t)$.

As for first order traffic models, we can define the demand and supply functions for each road i as follows: for a given constant c (corresponding to a fixed value of $v + p_i(u)$), we have

$$
D_i(u, c, t) = \begin{cases} \big(c - p_i(u, t)\big)u & \text{if } u \le \sigma_i(c, t), \\ \big(c - p_i(\sigma_i(c, t), t)\big)\sigma_i(c, t) & \text{if } u \ge \sigma_i(c, t), \end{cases} \tag{4.46}
$$

$$
S_i(u, c, t) = \begin{cases} \big(c - p_i(\sigma_i(c, t), t)\big)\sigma_i(c, t) & \text{if } u \le \sigma_i(c, t), \\ \big(c - p_i(u, t)\big)u & \text{if } u \ge \sigma_i(c, t), \end{cases} \tag{4.47}
$$

where

$$
\sigma_i(c, t) = u_i^{\max}\left(\frac{c\,\gamma_i}{v_i^{\mathrm{ref}}(t)\,(1 + \gamma_i)}\right)^{\frac{1}{\gamma_i}} \tag{4.48}
$$

is the sonic point on the curve $\{v + p_i(u, t) = c\}$ in the (u, uv)-plane. Figure 4.23 provides an illustration of the considered demand and supply functions.

4.5.1.1   Coupling and Boundary Conditions

We describe here the different type of junctions used later and the corresponding coupling conditions. The considered coupling and boundary conditions can be expressed in terms of mass flow $q = uv$ and "momentum flow" $q(v + p_i(\rho))$.

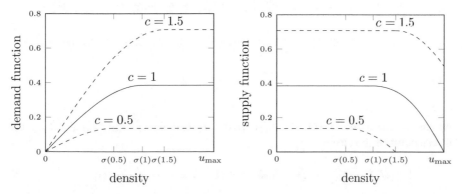

Fig. 4.23 Demand and supply functions corresponding to $u_{\max} = 1$, $v^{\mathrm{ref}} = 2$, $\gamma = 2$, and the given values for c

One-to-One Junction

In the case of one-to-one junction, the flow maximization over all admissible states leads to

$$q_1 = q_2 = \tilde{q} = \min\left\{D_1\big(u_1, v_1 + p_1(u_1, t), t\big), S_2\big(\tilde{u}_2, v_1 + p_1(u_1, t), t\big)\right\},$$
(4.49)

where $\tilde{u}_2$ is either obtained by the intersection of the curves

$$\{v_2(U, t) = v_2(U_2, t)\} \quad \text{and} \quad \{v_2 + p_2(U, t) = v_1 + p_1(u_1, t)\} \tag{4.50}$$

or $\tilde{u}_2 = 0$.

On-Ramp at a One-to-One Junction

Let us consider a one-to-one junction with an on-ramp, and the demand $D_2(l, t)$ at the on-ramp is computed by

$$D_2(l, t) = \omega_2(t) \begin{cases} f_2^{\max} & \text{if } l > 0, \\ \min\{f_2^{\text{in}}(t), f_2^{\max}\} & \text{if } l = 0, \end{cases} \tag{4.51}$$

where l is the length of the queue, $\omega_2(t) \in [0, 1]$ is the (time-dependent) metering rate, $f_2^{\text{in}}(t)$ is the "inflow" of cars arriving at the on-ramp, and $f_2^{\max}$ is the maximum flow onto the main road. To get a unique solution, we assign the priority parameter P to road 1 and set

$$q_1 = \min\left\{D_1\big(u_1, W_1(t), t\big), \max\left\{PS_3\big(\tilde{u}_3, W_1(t), t\big), S_3\big(\tilde{u}_3, W_1(t), t\big) - D_2\big(l_2, t\big)\right\}\right\},$$
(4.52)

$$q_2 = \min\left\{D_2\big(l_2, t\big), \max\left\{(1 - P)S_3\big(\tilde{u}_3, W_1(t), t\big), S_3\big(\tilde{u}_3, W_1(t), t\big) - D_1\big(u_1, W_1(t), t\big)\right\}\right\},$$
(4.53)

$$q_3 = q_1 + q_2, \tag{4.54}$$

where we have set $W_1(t) = v_1 + p_1(u_1, t)$ and $\tilde{u}_3$ is either obtained by the intersection of the curves

$$\{v_3(U, t) = v_3(U_3, t)\} \quad \text{and} \quad \{w_3(U, t) = v_1 + p_1(u_1, t)\} \tag{4.55}$$

or $\tilde{u}_3 = 0$. The evolution of the queue length at the on-ramp is governed by

$$\frac{d \, l_2(t)}{\Delta t} = f_2^{\text{in}}(t) - q_2, \tag{4.56}$$

where we may assume $l_2(0) = 0$.

On-Ramp at Origin

To better keep track of vehicles entering the network, we consider an on-ramp at vertexes with only one outgoing road:

$$D_1(l, t) = \omega_1(t) \begin{cases} f_1^{max} & \text{if } l > 0, \\ \min\{f_1^{in}(t), f_1^{max}\} & \text{if } l = 0 \end{cases} \tag{4.57}$$

for the demand at the on-ramp, where l is the length of the queue, $\omega_1(t) \in [0, 1]$ is the (time-dependent) metering rate, $f_1^{in}(t)$ is the inflow of cars arriving at the on-ramp, and f_1^{max} is the maximum flow onto the road.

Since in this case there is no value for c to evaluate the supply function of the road as in (4.49), we consider an auxiliary left state U_1 mimicking the desired inflow of the on-ramp. We assume that the velocity of the auxiliary state equals the desired velocity of the road, i.e.,

$$U_1 = \left(u, u\tilde{W}_2(u, t)\right) \quad \text{such that} \quad uV_2(u, t) = D_1(l, t), \tag{4.58}$$

where $\tilde{W}(u, t) = V_2(u, t) + p_2(u, t)$, which is solved by

$$u_{\pm} = \frac{u_{max}^2}{2} \pm \sqrt{\frac{(u_{max}^2)^2}{4} - \frac{u_{max}^2 D_1(l, t)}{v_{max}^2(t)}}. \tag{4.59}$$

The choice of u_- fulfills $D_2(u_-, \tilde{w}, t) = D_1(l, t)$.

Finally, we get

$$q_1 = q_2 = \tilde{q} = \min\left\{D_1(l_1, t), S_2\left(\tilde{u}_2, \tilde{W}_2(u_-, t), t\right)\right\} \tag{4.60}$$

with $\tilde{W}_2(u_-, t) = V_2(u_-, t) + p_2(u_-, t)$, and $\tilde{u}_2$ is either obtained by the intersection of the curves

$$\{v_2(U, t) = v_2(U_2, t)\} \quad \text{and} \quad \{w_2(U, t) = \tilde{W}_2(u_-, t)\} \tag{4.61}$$

or $\tilde{u}_2 = 0$.

The evolution of the queue at the on-ramp is given by

$$\frac{d\,l_1(t)}{\Delta t} = f_1^{in}(t) - q_1. \tag{4.62}$$

Outflow Conditions

At nodes with only one ongoing road, we consider absorbing boundary conditions up to a given maximum flow rate $f_1^{\text{out}}(t)$:

$$q_1 = \min\left\{D_1\big(u_1, v_1 + p_1(u_1, t), t\big),\ f_1^{\text{out}}(t)\right\}. \tag{4.63}$$

The momentum flow is given by $q_1(v_1 + p_1(u_1, t))$.

4.5.2 Numerical Simulations for Aw–Rascle on Network with Control

For the numerical solution of the described model, we consider a finite number of time points $t^n = n\Delta t$, where $\Delta t = T/Nt$. Moreover, each road i is divided into Nx_i uniform cells of size $\Delta x_i = L_i/Nx_i$. We also set $\Delta t\, v_{\max}^i \le \Delta x_i$ in all numerical simulations.

4.5.2.1 Numerical Method

System (4.43) is discretized via a fractional step method composed by a first order Godunov scheme for the flux term and an implicit Euler method for the relaxation term.

On each road i, the initial conditions (for $j \in \{1, \ldots, Nx_i\}$) are given by the cell averages

$$U_{i,j-0.5}^0 = \frac{1}{\Delta x_i} \int_{(j-1)\Delta x_i}^{j\Delta x_i} U_i(x, 0)\, \Delta x \qquad \text{for } j \in \{1, \ldots, Nx_i\}. \tag{4.64}$$

Then, at each time iteration $n \in \{0, \ldots, Nt - 1\}$, we compute

$$\tilde{U}_{i,j-0.5}^{n+1} = U_{i,j-0.5}^n - \frac{\Delta t}{\Delta x_i}\big(F_{i,j}^n - F_{i,j-1}^n\big), \qquad \text{(transport)} \tag{4.65}$$

$$U_{i,j-0.5}^{n+1} = \tilde{U}_{i,j-0.5}^{n+1} + \Delta t\, g(U_{i,j-0.5}^{n+1}). \qquad \text{(relaxation)} \tag{4.66}$$

The boundary fluxes $F_{i,0}^n$ and F_{i,Nx_i}^n are given via coupling/boundary conditions, while

$$F_{i,j}^n = \begin{pmatrix} q_{i,j}^n \\ (v_{i,j-0.5}^n + p_i(u_{i,j-0.5}^n, t^n))\, q_{i,j}^n \end{pmatrix} \qquad \text{for } j \in \{1, \ldots, Nx_i\}, \tag{4.67}$$

with

$$q_{i,j}^n = \min \left\{ D_i(u_{i,j-0.5}^n, w_{i,j-0.5}^n, t^n), \ S_i(\tilde{u}_{i,j}^n, w_{i,j-0.5}^n, t^n) \right\}, \qquad (4.68)$$

where $\tilde{u}_{i,j}^n$ is either obtained by the intersection of the curves

$$\{v_i(U, t^n) = v_{i,j+0.5}^n\} \quad \text{and} \quad \{v_i^t + p_i(U, t^n) = w_{i,j-0.5}^n\} \qquad (4.69)$$

or $\tilde{u}_{i,j}^n = 0$, and $w_{i,j-0.5}^n = v_{i,j-0.5}^n + p_i(u_{i,j-0.5}^n, t^n)$.

4.5.2.2 Numerical Results

We study two different scenarios:

1. The capacity drop for a one-to-one junction with on-ramp
2. Speed control and coordinated ramp metering for a simple network

4.5.2.3 Capacity Drop

We first focus the attention on the capacity drop effect, cf. [169]. We will show how, increasing the inflow into the network, the outflow will decrease.

We consider the network depicted in Fig. 4.24 consisting of two roads of 1 km length with an on-ramp in between. On both roads, we set $u_{max} = 180 \frac{\text{cars}}{\text{km}}$, $v_{max} = v^{\text{ref}} = 100 \frac{\text{km}}{\text{h}}$, $\gamma = 2$, $\delta = 0.005$ h, and an initial density of 50 $\frac{\text{cars}}{\text{km}}$. At the origin "in," we assume a constant (desired) inflow $f_1^{\text{in}} = 3500 \frac{\text{cars}}{\text{h}}$. At the on-ramp, we consider time-dependent inflows, starting from $f_2^{\text{in}} = 500 \frac{\text{cars}}{\text{h}}$ up to 2500 $\frac{\text{cars}}{\text{h}}$ and down to 500 $\frac{\text{cars}}{\text{h}}$ again. The priority parameter at the on-ramp is $P = 0.5$.

Table 4.7 and Fig. 4.25 show the simulation results for the AR and the LWR model (with $\Delta x = 100$ meters and $\Delta t = 1.8$ seconds), i.e., the resulting stationary states. The first two columns report the desired and the actual inflow at the on-ramp. The following three values are the resulting density, velocity, and the value of $v + p(u)$ just upstream the on-ramp (end of the first road). The last two columns show the total outflow at the end of the second road.

We observe that up to 1000 $\frac{\text{cars}}{\text{h}}$ of inflow at the on-ramp, the total inflow (of 4500 $\frac{\text{cars}}{\text{h}}$) is within the capacity of the second road. For higher values, the resulting total flux cannot be received at some point. Then the value of $v + p(u)$

Fig. 4.24 One-to-one junction with on-ramp

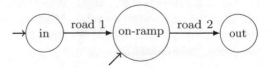

Table 4.7 Capacity drop effect

Inflow at on-ramp [$\frac{cars}{h}$]		u_1	v_1	$v_1 + p(u_1)$	Outflow AR	Outflow LWR
Desired	Actual	[$\frac{cars}{km}$]	[$\frac{km}{h}$]	[$\frac{km}{h}$]	[$\frac{cars}{h}$]	[$\frac{cars}{h}$]
500	500	47.6	73.6	77.1	4000	4000
1000	1000	47.6	73.6	77.1	4500	4500
1500	1500	156.4	13.1	50.9	3554	4500
2000	1764	160.2	11.0	50.6	3527	4500
2500	1764	160.2	11.0	50.6	3527	4500
1000	1000	148.0	17.8	51.6	3629	4500
500	500	137.2	23.8	52.8	3762	4000

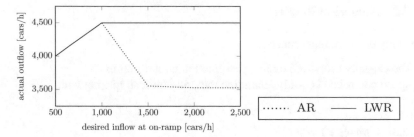

Fig. 4.25 Actual outflow depending on the desired inflow at the on-ramp for the AR and the LWR model

at the end of the first road impacts the total flux entering the second road. As a consequence, the outflow at the end of the second road for the cases $f_2^{in} \in$ {1500 $\frac{cars}{h}$, 2000 $\frac{cars}{h}$, 2500 $\frac{cars}{h}$} is lower than the outflow for the cases $f_2^{in} \in$ {500 $\frac{cars}{h}$, 1000 $\frac{cars}{h}$}. Due to the choice of the priority parameter $P = 0.5$, the effect of a decreasing outflow while the desired inflow increases stagnates as soon as the fluxes from the first road and the on-ramp are equal, which is the case for $f_2^{in} \in$ {2000 $\frac{cars}{h}$, 2500 $\frac{cars}{h}$}.

We point out that, even when the inflow at the on-ramp is lowered again, the original state for the same values of f_1^{in} and f_2^{in} is not reached. This is due to the fact that the outflow in the congested situation is below the accumulated desired inflows. Therefore, the queue at the origin keeps increasing even though the capacity of the road could handle the desired inflows in the free flow situation.

Considering the same scenario with the LWR model (Sect. 4.4 for details), we see that the outflow increases with the accumulated desired inflows until it reaches the maximum capacity of 4500 $\frac{cars}{h} = \frac{u_{max}}{2} \cdot \frac{v_{max}}{2}$ (see again Table 4.7 and Fig. 4.25). According to the chosen priority $P = 0.5$, the actual inflows at the origin and at the on-ramp are 2250 $\frac{cars}{h}$ at this stage. Further increasing the inflow at the on-ramp does not affect the situation on the roads. Moreover, unlike the AR model, decreasing the desired inflow f_2^{in} below 1000 $\frac{cars}{h}$ leads back to the original situation (as soon as the queues have emptied).

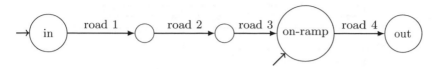

Fig. 4.26 Road network with an on-ramp

Table 4.8 Parameters of the roads in Fig. 4.26

Road	Length [km]	u_{max} [$\frac{cars}{km}$]	v^{low} [$\frac{km}{h}$]	v^{high} [$\frac{km}{h}$]	Initial density [$\frac{cars}{km}$]
1	2	180	100	100	50
2	1	180	50	100	50
3	1	180	50	100	50
4	2	180	100	100	50

4.5.2.4 Coordinated Speed Control and Ramp Metering

In this section, we aim at controlling maximal speed limits and on-ramp inflows to minimize the total travel time. The ramp metering rate $\omega_i(t)$ in Eqs. (4.51) and (4.57) is used to control the demand at the on-ramp. As in Sect. 4.4, we assume $v^i_{max} = v^i_{max}(t) \in [v^{low}_i, v^{high}_i]$ is another (piecewise constant) control variable. Regarding the pressure term, we will consider two variants:

1. $v^{ref}_i(t) = v^i_{max}(t)$, i.e., the pressure term directly depends on the (controllable) speed limit and therefore $p_i(u) = p_i(u, t)$.
2. $v^{ref}_i(t) = v^{high}_i$, i.e., the pressure term is independent of the current speed limit.

We consider the road network represented in Fig. 4.26 (adapted from [174]) with the road parameters given in Table 4.8. The exponents in the pressure function are $\gamma_i - \gamma = 2$ for all roads and $\delta = 0.005$ h for the relaxation parameter. The priority parameter at the on-ramp is $P = 0.5$ and $f^{max} = 2000 \frac{cars}{h}$. At the origin "in," we consider $f^{max} = 4000 \frac{cars}{h}$.

We fix a time horizon of $T = 3.0$ hours and the boundary conditions depicted in Fig. 4.27. We aim at minimizing the total travel time

$$\sum_{i=1}^{4} \int_0^T \int_0^{L_i} u_i(x, t) \, \Delta x \, \Delta t + \sum_{j=1}^{2} \int_0^T l_j(t) \, \Delta t, \qquad (4.70)$$

given an upper bound of 100 vehicles in the queue of the on-ramp. To solve this optimization task, we apply a first-discretize-then-optimize approach and adjoint calculus.

Table 4.9 shows the total travel time for the AR and LWR models with and without optimal control. The resulting queues for the AR model with scaling of the pressure function ($\frac{\partial p_i}{\partial v^i_{max}} \neq 0$) are shown in Fig. 4.28. Without control, there is

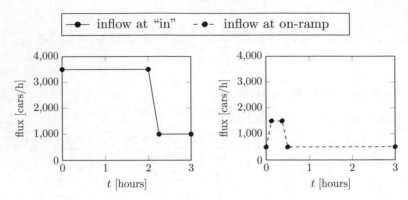

Fig. 4.27 Inflow profiles for the network in Fig. 4.26

Table 4.9 Optimization results for the network in Fig. 4.26

	AR, $\frac{\partial p_i}{\partial v_{max}^i} \neq 0$	AR, $\frac{\partial p_i}{\partial v_{max}^i} = 0$	LWR
No control	1871.7	1871.7	834.9
Ramp metering only	1325.3	1325.3	834.9
Speed control only	1122.8	872.6	834.9
Both control types	814.5	818.4	834.9

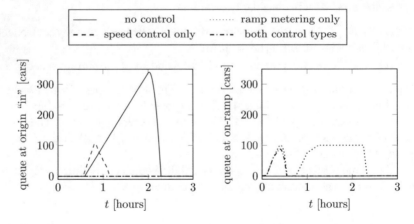

Fig. 4.28 Queue at the origin "in" and the on-ramp with and without optimal control

no queue at the on-ramp, while more than 300 cars accumulate in the queue at the origin. With ramp metering alone, the queue at the origin is reduced to zero, and up to 100 cars accumulate in the queue at the on-ramp. When ramp metering and speed control are both implemented, the upper bound of 100 cars is never reached. In the case of speed control alone, we have no queue at the on-ramp, while some cars accumulate at the origin.

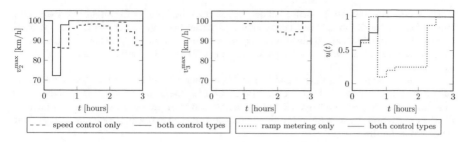

Fig. 4.29 Optimal control of $v_{max}^i(t)$ on road 2 (left) and road 3 (center) and $\omega(t)$ at the on-ramp (right)

For the LWR model, both queues stay empty during the whole time horizon even without controls, so the optimization produces the same result. This is due to the inability of the LWR model to capture the capacity drop effect.

Figure 4.29 shows the applied optimal controls.

4.6 Bibliographical Notes

The problem of distributed control for partial differential equations with application to traffic has been addressed in several papers in the literature.

An overview of the variable speed limit approaches can be found in [191]. The first papers dealing with variable speed limit using a discretized system of hyperbolic partial differential equations date back to the end of the nineties [3, 214]. In particular, the work [214] used the speed as control to stabilize the nonlinear hyperbolic PDE even with disturbances in the initial data. In the engineering literature, several papers deal with variable speed limit by controlling a discretized hyperbolic PDE [32, 63, 173, 227, 238]. The control is the speed limit that is used to detect and dissipate shock waves generated by macroscopic traffic models. In [173], the problem is analyzed following a Model Predictive Control approach, where the optimization is obtained using sequential quadratic programming (SPQ), while in [63, 96] the problem is analyzed by using closed loop/feedback approaches. For an overview of the approaches on the continuous hyperbolic partial differential equation, in addition to the papers used in this chapter [116, 153, 193] recently, the work [186] looked at the 1D scalar partial differential equation

$$\partial_t u + \partial_x(v(t, x)u(t, x)) = 0, \tag{4.71}$$

with VSL applied continuously in time and space along a freeway.

In the last few years, variable speed limit approaches have been used also in the case of moving actuators. In particular, autonomous vehicles can be used to control traffic by adopting specific speed policies, see for example [31, 167].

Chapter 5
Lagrangian Control of Conservation Laws and Mixed Models

5.1 Introduction

A vehicle with different (eventually controlled) dynamics from general traffic along a street may reduce the road capacity, thus generating a *moving bottleneck*, and can be used to act on the traffic flow. The interaction between the controlled vehicle and the surrounding traffic, and the consequent flow reduction at the bottleneck position, can be described either by a conservation law with space dependent flux function [200], or by a strongly coupled PDE-ODE system proposed in [112, 208].

Many researchers addressed the problem of detecting a single vehicle among traffic. Most of these are PDE-ODE models where the main traffic is described by a scalar conservation law and the single vehicle's trajectory satisfies an ODE. In these early works, see [81, 82], the car did not influence the surrounding traffic, making the system not fully coupled. More complicated situations involve cases in which the car influences in turn the bulk traffic. These cases give rise to more interesting control problems, where it is possible to control the main traffic stream by acting on the dynamics of single (possibly many) vehicles among the fleet. The main idea is to create moving bottlenecks that produce a non-negligible capacity drop in the main traffic flow.

The problem of fixed and moving bottlenecks was first introduced in the engineering literature [101, 102, 203, 208–210]. In the mathematical literature the first models involving fixed bottlenecks were proposed in [78] and [140]. The model in [78] considers the case of a tollgate that is modeled as a pointwise constraint on the maximal flux. This model can be used for control purposes as showed in Sect. 3.5. The second paper uses one-to-one junctions to account for different capacities. Later on the problem of moving bottlenecks was addressed in [112, 200]. Ad-hoc numerical schemes were introduced to solve these models numerically, see [67, 111, 146]. Moreover, extension to second order models or vehicle platooning were developed in [137, 243] respectively. These models have been used in the latest years to address control problems for traffic by using ad-hoc capacity drops, variable

A. Bayen et al., *Control Problems for Conservation Laws with Traffic Applications*, PNLDE Subseries in Control 99, https://doi.org/10.1007/978-3-030-93015-8_5

speed limits and autonomous vehicles (AVs). In fact, it is possible to describe the dynamics of a controller involving variable speed limits and/or pointwise capacity drops as a moving bottlenecks, see [167, 181, 228, 246].

5.2   PDE-ODE Models for Moving Bottlenecks

As usual, the flux function f is given by $f(u) = uv(u)$, where $v = v(u)$ is the average speed of cars, with $v(0) = v_{max}$ and $v(u_{max}) = 0$. We assume that the flux satisfies the condition

(F) $f : \mathbf{C}^2 ([0, u_{max}]; [0, +\infty))$, $f(0) = f(u_{max}) = 0$,
   f strictly concave: $-B \leq f''(u) \leq -\beta < 0$ for all $u \in [0, u_{max}]$, for some
   $\beta, B > 0$.

We note that **(F)** implies that $v'(\rho) < 0$ for every $\rho \in]0, R[$, see [128, Lemma 1].

5.2.1   A Macroscopic Model with Space Dependent Flux

In [146, 200], the authors introduce a macroscopic model based on the LWR traffic flow model [224, 249], in which the capacity drops due to the controlled vehicle is modeled by a smooth cut-off function multiplying the flux. More precisely, the model reads as follows:

$$\partial_t u + \partial_x f_\varphi(x, y(t), u) = 0, \qquad\qquad t > 0, \ x \in \mathbb{R}, \qquad (5.1a)$$

$$\dot{y}(t) = \omega v(u(t, y(t))), \qquad\qquad t > 0, \qquad (5.1b)$$

$$u(0, x) = u_0(x), \qquad\qquad x \in \mathbb{R}, \qquad (5.1c)$$

$$y(0) = y_0, \qquad\qquad (5.1d)$$

where $u = u(t, x) \in [0, u_{max}]$ is the density of cars, $y = y(t)$ is the position of the controlled vehicle, $\omega \in]0, \kappa[$ is its maximal speed for some $\kappa \in]0, 1]$, and the flux function f_φ is given by

$$f_\varphi(x, y, u) = f(u) \cdot \varphi(x - y). \qquad (5.2)$$

In (5.2), $\varphi(\xi)$ is a function representing the capacity dropping of car flows, due to the presence of the slower vehicle. We assume further the cut–off function $\varphi(\xi)$ to be smooth and that there exist $0 < \kappa < 1$ and $\delta > 0$ such that

$(\varphi.1)$ $\kappa \leq \varphi(\xi) \leq 1$, for every $\xi \in \mathbb{R}$;
$(\varphi.2)$ $\varphi(\xi) = 1$ for every $\xi \notin [-\delta, \delta]$;
$(\varphi.3)$ $\varphi(0) = \kappa$;
$(\varphi.4)$ φ is strictly decreasing in $] - \delta, 0[$ and strictly increasing in $]0, \delta[$;

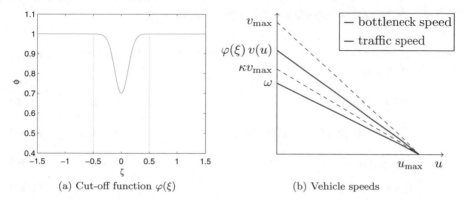

(a) Cut-off function $\varphi(\xi)$ (b) Vehicle speeds

Fig. 5.1 (a) An example of the cut-off function $\varphi(\xi)$ with $\delta = 0.5$ and $\kappa = 0.7$. (b) Moving bottleneck speed $\omega\, v(u)$ and mean traffic speed at $\xi = x - y(t)$

see Fig. 5.1a. Observe that the assumption $\omega < \kappa$ ensures that the other vehicles can overtake the moving bottleneck caused by the controlled one.

The solutions to (5.1) are intended in the usual weak sense.

Definition 21 A vector (u, y) is a solution to (5.1) if $u(t, \cdot)$ is a function in $\mathbf{BV}(\mathbb{R})$ for a.e. $t > 0$ which solves (5.1a) in the sense of distributions, that is

$$\int_0^{+\infty} \int_{\mathbb{R}} \{u\, \partial_t\, \phi(t, x) + f(x, y(t), u)\, \partial_x\, \phi(t, x)\}dxdt + \int_{\mathbb{R}} u_0(x)\phi(0, x)dx = 0,$$

for any $\phi \in \mathbf{C}_c^1(\mathbb{R}^2; \mathbb{R}^+)$.

Moreover, the position of the slower moving vehicle $y(t)$ solves equation (5.1b) in $[0, T]$ in the sense of Filippov (see [132]), namely $y(t)$ is an absolutely continuous function such that $y(0) = y_0$ and

$$\dot{y}(t) \in \overline{\mathrm{co}}\{\omega\, v(u) : u \in \mathcal{I}[u(t, y(t)-), u(t, y(t)+)]\}$$

for a.e. $t > 0$, where the set $\mathcal{I}[a, b]$ is defined as the smallest interval containing a and b.

Existence of such solutions is proven in [200] for any $u_0 \in \mathbf{BV}(\mathbb{R})$. The proof relies on the construction of approximate solutions by a fractional step method. A wave-front tracking approach can be found in [56]. Multiple bottlenecks can be handled as in [56, Section 6] or [146].

5.2.2   PDE-ODE Models with Flux Constraint

Following the model proposed in [112, 138, 208], we consider the following coupled PDE-ODE system

$$\partial_t u + \partial_x f(u) = 0, \qquad\qquad t > 0, \; x \in \mathbb{R}, \qquad (5.3a)$$

$$\dot{y}(t) = \min\{\omega(t), v(u(t, y(t)+))\}, \qquad\qquad t > 0, \qquad (5.3b)$$

$$f(u(t, y(t))) - \dot{y}(t)u(t, y(t)) \le F_\alpha(\dot{y}(t)), \qquad\qquad t > 0, \qquad (5.3c)$$

$$u(0, x) = u_0(x), \qquad\qquad x \in \mathbb{R}, \qquad (5.3d)$$

$$y(0) = y_0, \qquad\qquad (5.3e)$$

where u_0 and y_0 are the initial traffic density and the controlled vehicle position, and $\omega(t) \in [0, v_{\max}]$ is the desired speed of the controlled vehicle at time $t > 0$.

Remark 12 Unlike the previous model (5.1b), in (5.3b) the controlled vehicle travels at its own speed $\omega(t)$ if downstream traffic conditions allow it, i.e., if $v(u(t, y(t)+)) \ge \omega(t)$, that is $u(t, y(t)+) \le v^{-1}(\omega(t)) =: u^*(t)$, see Fig. 5.2.

The function F_α in (5.3c) is defined as

$$F_\alpha(\dot{y}(t)) := \max_{u \in [0, u_{\max}]} (\alpha f(u/\alpha) - u\dot{y}(t)), \qquad \alpha \in \,]0, 1[, \qquad (5.4)$$

and represents the road capacity reduction due to the presence of the vehicle, acting as a moving bottleneck which imposes a unilateral flux constraint at its position. The coefficient α is the capacity reduction rate given by the ratio of the number of lanes non occupied by the vehicle over the total number of lanes.

Fig. 5.2 The speed of the moving bottleneck: $u^*(t)$ is defined by the equality $v(u^*(t)) = \omega(t)$

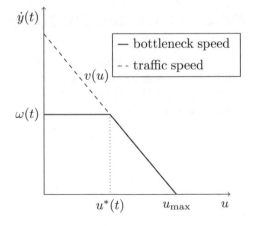

To determine the function F_α, we consider reduced flux function

$$f_\alpha : [0, \alpha u_{\max}] \longrightarrow \mathbb{R}^+$$
$$u \longmapsto uv(u/\alpha) = \alpha f(u/\alpha),$$

which is a strictly concave function satisfying $f_\alpha(0) = f_\alpha(\alpha u_{\max}) = 0$. For every $\omega \in [0, v_{\max}]$, define the point $\tilde{u}_\omega$ as the unique solution to the equation $f_\alpha'(u) = u$. Introduce also, for every $\omega \in [0, v_{\max}]$, the function

$$\varphi_\omega : [0, u_{\max}] \longrightarrow \mathbb{R}^+$$
$$u \longmapsto f_\alpha(\tilde{u}_\omega) + \omega(u - \tilde{u}_\omega).$$

Hence, if $\dot{y}(t) = \omega$, the function F_α in (5.3c) is defined by

$$F_\alpha : [0, v_{\max}] \longrightarrow \mathbb{R}^+$$
$$\omega \longmapsto \varphi_\omega(0) = f_\alpha(\tilde{u}_\omega) - \omega\tilde{u}_\omega. \tag{5.5}$$

If $\dot{y}(t) = v(u(t, y(t)+))$, the inequality (5.3c) is trivially satisfied since the left-hand side is zero. Finally, the points $0 \leq \check{u}_\omega \leq \tilde{u}_\omega \leq \hat{u}_\omega \leq u_\omega \leq u_{\max}$ are uniquely defined by

$$\check{u}_\omega = \min \mathcal{I}_\omega, \quad \hat{u}_\omega = \max \mathcal{I}_\omega, \quad \mathcal{I}_\omega = \{u \in [0, u_{\max}] : f(u) = \varphi_\omega(u)\},$$

and implicitly by

$$v(u_\omega) = \omega,$$

see [112, 138] and Fig. 5.3. It is straightforward to see that $\check{u}_{v_{\max}} = \tilde{u}_{v_{\max}} = \hat{u}_{v_{\max}} = u_{v_{\max}} = 0$.

Solutions of the Cauchy problem (5.3) are intended in the following weak sense.

Definition 22 A couple $(u, y) \in \mathbf{C}^0\left(\mathbb{R}^+; \mathbf{L}^1_{\text{loc}}(\mathbb{R}; [0, u_{\max}])\right) \times \mathbf{W}^{1,1}_{\text{loc}}(\mathbb{R}^+; \mathbb{R})$, with $TV(u(t, \cdot)) < +\infty$ for all $t \in \mathbb{R}^+$, is a solution to (5.3) if

Fig. 5.3 The definition of $\tilde{u}_\omega$, $\check{u}_\omega$, $\hat{u}_\omega$ and u_ω

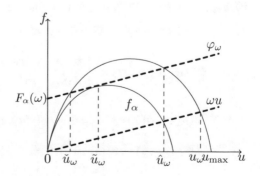

1. For every $\kappa \in \mathbb{R}$ and for all $\varphi \in \mathbf{C}_c^1(\mathbb{R}^2; \mathbb{R}^+)$ it holds

$$
\int_{\mathbb{R}^+} \int_{\mathbb{R}} (|u - \kappa| \, \partial_t \varphi + \mathrm{sgn}(u - \kappa) \, (f(u) - f(\kappa)) \, \partial_x \varphi) \, \mathrm{d}x \, \mathrm{d}t
$$

$$
+ \int_{\mathbb{R}} |u_0 - \kappa| \varphi(0, x) \, \mathrm{d}x \tag{5.6a}
$$

$$
+ 2 \int_{\mathbb{R}^+} (f(\kappa) - \dot{y}(t)\kappa - \min\{f(\kappa) - \dot{y}(t)\kappa, F_\alpha(\dot{y}(t))\}) \, \varphi(t, y(t)) \, \mathrm{d}t \geq 0;
$$

2. y is a Carathéodory solution of (5.3b), (5.3e), i.e., for a.e. $t \in \mathbb{R}^+$

$$
y(t) = y_0 + \int_0^t \min\{\omega(t), v(u(s, y(s)+))\} \, \mathrm{d}s; \tag{5.6b}
$$

3. the constraint (5.3c) is satisfied, in the sense that for a.e. $t \in \mathbb{R}^+$

$$
\lim_{x \to y(t)\pm} (f(u) - u \min\{\omega(t), v(u)\}) \, (t, x) \leq F_\alpha(\dot{y}(t)), \tag{5.6c}
$$

with F_α defined in (5.4).

Well-posedness results for (5.3) are based on wave-front tracking approximations. Therefore, we focus here on the solution to the Riemann problem, i.e., problem (5.3) with initial data

$$
y_0 = 0 \quad \text{and} \quad u_0(x) = \begin{cases} u_L & \text{if } x < 0, \\ u_R & \text{if } x > 0. \end{cases} \tag{5.7}
$$

Denote by $\mathcal{R}$ the standard (i.e., without the constraint (5.3c)) Riemann solver for (5.3a)–(5.7), i.e., the (right continuous) map $(t, x) \mapsto \mathcal{R}(u_L, u_R)(x/t)$ given by the standard weak entropy solution, see Sect. A.6.1. Following [112, Definition 3.1], the constrained Riemann solver is defined below.

Definition 23 For any $\omega \in [0, v_{\max}]$, the constrained Riemann solver $\mathcal{R}^\omega :$ $[0, u_{\max}]^2 \to \mathbf{L}_{loc}^1(\mathbb{R}; [0, u_{\max}])$ is defined as follows.

1. If $f(\mathcal{R}(u_L, u_R)(\omega)) > F_\alpha(\omega) + \omega \mathcal{R}(u_L, u_R)(\omega)$, then

$$
\mathcal{R}^\omega(u_L, u_R)(x/t) = \begin{cases} \mathcal{R}(u_L, \hat{u}_\omega)(x/t) & \text{if } x < \omega t, \\ \mathcal{R}(\check{u}_\omega, u_R)(x/t) & \text{if } x \geq \omega t, \end{cases} \quad \text{and} \quad y(t) = \omega t.
$$

2. If $f(\mathcal{R}(u_L, u_R)(\omega)) \leq F_\alpha(\omega) + \omega \mathcal{R}(u_L, u_R)(\omega)$, then

$$
\mathcal{R}^\omega(u_L, u_R) = \mathcal{R}(u_L, u_R) \quad \text{and} \quad y(t) = \min\{\omega, v(u(t, y(t)+))\} \, t.
$$

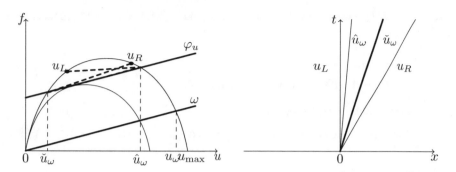

Fig. 5.4 The constrained Riemann problem (5.3)–(5.7), case 1. In this case $y(t) = y(0) + \omega t$, while the solution u is composed by two classical shocks separated by an undercompressive shock between $\hat{u}_\omega$ and $\check{u}_\omega$

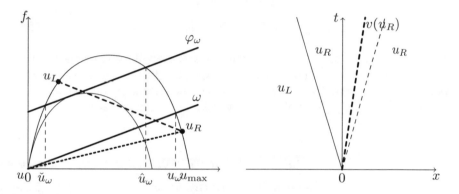

Fig. 5.5 The constrained Riemann problem (5.3)–(5.7), case 2. In this case $y(t) = y(0) + v(u_R)t$, while the solution for u is composed by the classical shock connecting u_L to u_R

Figures 5.4 and 5.5 illustrate two possible configurations corresponding to points 1. and 2. in Definition 23, respectively. In particular, we remark that when the constraint is enforced (point 1.) the jump discontinuity from $\hat{u}_\omega$ to $\check{u}_\omega$ is an *undercompressive shock*, which satisfies Rankine–Hugoniot conditions but violates Lax entropy conditions, which motivates the last term in (5.6a).

We finally have the following general existence result [138], holding for initial data and controls with bounded total variation.

Theorem 32 *Let the initial conditions $u_0 \in \mathbf{BV}\left(\mathbb{R}; [0, u_{\max}]\right)$, $y_0 \in \mathbb{R}$, and the open-loop control $\omega \in \mathbf{BV}\left(\mathbb{R}^+; [0, v_{\max}]\right)$. Then there exists a solution (u, y) to (5.3) in the sense of Definition 22.*

For partial stability results, see [114, 223].

The construction presented above can be extended to second order models, consisting in 2×2 systems of conservation laws accounting for mass conservation and an acceleration balance equation. In [272], the Aw–Rascle–Zhang (ARZ) model [23, 277] is coupled with a moving bottleneck (with constant maximal speed) as follows:

$$\begin{cases} \partial_t u + \partial_x (uv) = 0, \\ \partial_t (u(v + p(u))) + \partial_x (uv(v + p(u))) = 0, \end{cases} \qquad x \in \mathbb{R},\ t > 0, \qquad (5.8a)$$

$$\dot{y}(t) = \min \{\omega, v(t, y(t)+)\}, \qquad (5.8b)$$

$$\lim_{x \to y(t)\pm} (u(v - \min\{\omega, v\})) (t, x) \le F_\alpha(\dot{y}(t)), \qquad (5.8c)$$

with initial conditions

$$u(0, x) = u_0(x), \qquad (5.9a)$$

$$v(0, x) = v_0(x), \qquad (5.9b)$$

$$y(0) = y_0. \qquad (5.9c)$$

Above, $u = u(t, x)$ and $v = v(t, x)$ denote respectively the density and the mean velocity of traffic. Note that the quantity $v + p(u)$, usually referred to as *Lagrangian marker*, is transported at velocity v and depends on the density through a pressure law $p \in \mathbf{C}^2([0, +\infty[; [0, +\infty[)$ satisfying the following hypotheses:

$$p(0) = 0, \qquad (5.10a)$$

$$p'(u) > 0 \quad \text{for every } u > 0, \qquad (5.10b)$$

$$u \mapsto up(u) \quad \text{is strictly convex.} \qquad (5.10c)$$

The maximal speed ω of the controlled vehicle satisfies $0 < \omega < v_{\max}$. Hence, the phase space for (5.8a) is defined by the domain

$$\mathcal{D} = \{(u, v) \in \mathbb{R}^+ \times \mathbb{R}^+ : 0 \le v \le v_{\max}, \ 0 \le v + p(u) \le p(u_{\max})\}$$

away from the controlled vehicle position, and

$$\mathcal{D}_\alpha = \{(u, v) \in \mathbb{R}^+ \times \mathbb{R}^+ : 0 \le v \le v_{\max}, \ 0 \le v + p(u) \le p(\alpha u_{\max})\}$$

at $x = y(t)$.

When the controlled vehicle acts on the flow, namely when $v > \omega$ at $x = y(t)$, the left and right initial states of the Riemann problem associated with (5.8) will be joined by simple waves associated with the two characteristic fields, but the phase space $\mathcal{D}_\alpha$ must be preserved, namely that $0 \le v \le v_{\max}$ and $v + p(u) \le p(\alpha u_{\max})$ at the controlled vehicle position. The first constraint is already satisfied when the vehicle is not present. The second inequality is equivalent to

$$u(v - \omega) \le u (p(\alpha u_{\max}) - p(u) - \omega).$$

Under this form, it is clear that the constraint reads as a constraint on the relative flux at the bus position. Let us determine under which condition the quantity $v + p(u)$ equals a constant K such that $K \leq p(\alpha u_{\max})$, or equivalently

$$u(v - \omega) = u(K - p(u)) - u\omega,$$

with K such that $K \leq p(\alpha u_{\max})$. The function $u \to \phi(u; K) := u(K - p(u)) - u\omega$ is strictly concave in u by assumption (5.10c) and non-decreasing in K since $u \geq 0$. Moreover, we have $\phi(0; K) = 0$ for all K. Therefore, the maximal possible value F_α of the relative flux corresponds to $K = p(\alpha u_{\max})$ and

$$F_\alpha(\omega) = \max_{u \in [0, u_{\max}]} u \left(p(\alpha u_{\max}) - p(u) - \omega \right), \tag{5.11}$$

which is attained at u_α such that

$$p(\alpha u_{\max}) - u_\alpha p'(u_\alpha) - p(u_\alpha) - \omega = 0. \tag{5.12}$$

Note that we have to assume $\omega < p(\alpha u_{\max})$ to have $u_\alpha > 0$. From (5.11) and (5.12), we find that the largest admissible flux is given by $F_\alpha(\omega) = u_\alpha^2 p'(u_\alpha)$. Therefore, the classical solution will remain admissible provided that the relative flux does not exceed the upper bound $F_\alpha(\omega)$, see Fig. 5.6, where we have noted $w_\alpha := p(\alpha R)$ and $w_{max} := p(R)$. This criterion is the key ingredient to determine the two possible Riemann solutions described in [272].

Let $U_L = (u_L, v_L)$ and $U_R = (u_R, v_R)$ be two points in the domain $\mathcal{D}$. We consider the Riemann problem for (5.8) corresponding to the initial data

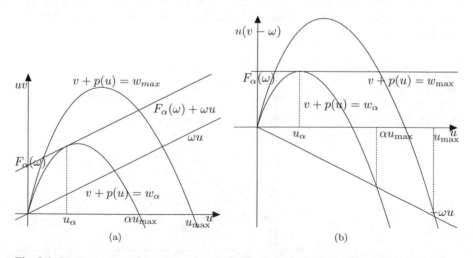

Fig. 5.6 Representation of the phase plane in the fixed and in the bus reference frames. (a) Flux representation in the phase plane. (b) Flux representation in the bus reference frame

$$u(0, x) = \begin{cases} u_L & \text{if } x < 0, \\ u_R & \text{if } x > 0, \end{cases} \tag{5.13a}$$

$$v(0, x) = \begin{cases} v_L & \text{if } x < 0, \\ v_R & \text{if } x > 0, \end{cases} \tag{5.13b}$$

$$y(0) = 0. \tag{5.13c}$$

and to the constant bottleneck speed $\dot{y}(t) = \omega$ for all $t > 0$. Let I be the set

$$I = \{u \in [0, u_{\max}] : u\,(v_L + p(u_L) - p(u)) = F_\alpha(\omega) + u\omega\}.$$

Since the map $u \mapsto u\,(v_L + p(u_L) - \omega - p(u))$ is strictly concave due to (5.10c), the set I contains at most two elements. If $I \neq \emptyset$, let $(\hat{u}, \hat{v})$ and $(\check{u}_1, \check{v}_1)$ be the points defined by

$$\hat{u} = \max I, \quad \hat{v} = \frac{F_\alpha(\omega)}{\hat{u}} + \omega, \quad \check{u}_1 = \min I \text{ and } \check{v}_1 = \frac{F_\alpha(\omega)}{\check{u}_1} + \omega.$$

These are respectively the points with maximal and minimal density of the Lax curve of the first family passing through (u_L, v_L) which satisfy the condition (5.8c) on the flux. Moreover, if $v_R > \omega$, we define the point $(\check{u}_2, \check{v}_2)$ as

$$\check{u}_2 = \frac{F_\alpha(\omega)}{v_R - \omega} \text{ and } \check{v}_2 = v_R.$$

This is the point of maximal density of the Lax curve of the second family passing through (u_R, v_R) for which (5.8c) is satisfied. All these points are depicted in Fig. 5.7.

Fig. 5.7 Notations used in the definition of the constrained Riemann solvers $\mathcal{R}^\omega_{ARZ1}$ and $\mathcal{R}^\omega_{ARZ2}$

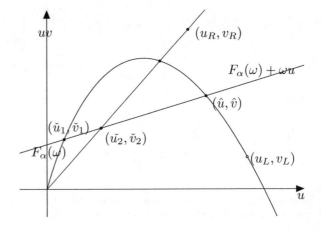

Let $\mathcal{R}_{ARZ}$ be the standard Riemann solver for (5.8a), (5.13a), (5.13b), see [23], and let

$$\bar{u}(U_L, U_R)(\cdot) \text{ and } \bar{v}(U_L, U_R)(\cdot)$$

be respectively the u and v components of $\mathcal{R}_{ARZ}(U_L, U_R)(\cdot)$, and let

$$f_1(\mathcal{R}_{ARZ}(U_L, U_R)(\cdot)) := \bar{u}(U_L, U_R)(\cdot)\, \bar{v}(U_L, U_R)(\cdot)$$

be the first component of the flux function of the ARZ system.

We propose now two possible definitions of Riemann solver for the constrained ARZ system (5.8).

Definition 24 The Riemann solver

$$\mathcal{R}^{\omega}_{ARZ1} : \mathcal{D} \times \mathcal{D} \to \mathbf{L}^1(\mathbb{R}, \mathbb{R}^+ \times \mathbb{R}^+)$$

is defined as follows.

1. If $f_1(\mathcal{R}_{ARZ}(U_L, U_R)(\omega)) > F_\alpha(\omega) + \omega\, \bar{u}(U_L, U_R)(\omega)$, then

$$\mathcal{R}^{\omega}_{ARZ1}(U_L, U_R)(x/t) = \begin{cases} \mathcal{R}_{ARZ}((u_L, v_L), (\hat{u}, \hat{v}))(x/t) & \text{if } x < y(t), \\ \mathcal{R}_{ARZ}((\check{u}_1, \check{v}_1), (u_R, v_R))(x/t) & \text{if } x > y(t), \end{cases}$$

and $y(t) = \omega t$.

2. If $f_1(\mathcal{R}_{ARZ}(U_L, U_R)(\omega)) \leq F_\alpha(\omega) + \omega\, \bar{u}(U_L, U_R)(\omega)$ then

$$\mathcal{R}^{\omega}_{ARZ1}(U_L, U_R)(x/t) = \mathcal{R}_{ARZ}(U_L, U_R)(x/t)$$

and $y(t) = \min\{\omega, \bar{v}(U_L, U_R)(\omega)\}\, t$.

Remark that the solution $\mathcal{R}^{\omega}_{ARZ1}$ is conservative for both density and momentum of the vehicles. Moreover, in case 1., the solution given by $\mathcal{R}^{\omega}_{ARZ1}$ does not satisfy the Lax entropy condition at the jump discontinuity between the left state $(\hat{u}, \hat{v})$ and the right state $(\check{u}_1, \check{v}_1)$, because $\check{u}_1 < \hat{u}$. Therefore, as in the scalar case, $(\hat{u}, \hat{v})$ and $(\check{u}_1, \check{v}_1)$ are connected by a *non-classical shock* (we refer the reader to [213] for an extensive survey on entropy-violating jump discontinuities).

Definition 25 The Riemann solver

$$\mathcal{R}^{\omega}_{ARZ2} : \mathcal{D} \times \mathcal{D} \to \mathbf{L}^1(\mathbb{R}, \mathbb{R}^+ \times \mathbb{R}^+)$$

is defined as follows.

1. If $f_1(\mathcal{R}_{ARZ}(U_L, U_R)(\omega)) > F_\alpha(\omega) + \omega\, \bar{u}(U_L, U_R)(\omega)$, then

$$\mathcal{R}^\omega_{ARZ2}(U_L, U_R)(x/t) = \begin{cases} \mathcal{R}_{ARZ}((u_L, v_L), (\hat{u}, \hat{v}))(x/t) & \text{if } x < y(t), \\ \mathcal{R}_{ARZ}((\check{u}_1, \check{v}_2), (u_R, v_R))(x/t) & \text{if } x > y(t), \end{cases}$$

and $y(t) = \omega t$.

2. If $f_1(\mathcal{R}_{ARZ}(U_L, U_R)(\omega)) \leq F_\alpha(\omega) + \omega\, \bar{u}(U_L, U_R)(\omega)$ then

$$\mathcal{R}^\omega_{ARZ2}(U_L, U_R)(x/t) = \mathcal{R}_{ARZ}(U_L, U_R)(x/t)$$

and $y(t) = \min\{\omega, \bar{v}(U_L, U_R)(\omega)\}\, t$.

The Riemann solver $\mathcal{R}^\omega_{ARZ2}$ conserves only the number of the vehicles. Indeed, along the line $x = y(t)$ the Rankine–Hugoniot condition holds for the first flux component, because both $(\hat{u}, \hat{v})$ and $(\check{u}_2, \check{v}_2)$ belong to the line $uv = F_\alpha(\omega) + \omega u$, but not for the second component.

Figure 5.8 presents a comparison of the solutions given by the Riemann solvers $\mathcal{R}^\omega_{ARZ1}$ and $\mathcal{R}^\omega_{ARZ2}$. We observe that the solver $\mathcal{R}^\omega_{ARZ1}$ gives a constant low density region just downstream the bus position, while applying $\mathcal{R}^\omega_{ARZ2}$ results in turn in a region in which the density is higher and non-constant. In particular, in the latter case the bus impacts the downstream traffic on a wider region.

5.2.3   A PDE-ODE Model for Vehicle Platooning

The first order LWR model can be adapted to describe the dynamics of the bulk traffic interacting with a platoon of vehicles. We denote by $z_D = z_D(t)$ and $z_U = z_U(t)$ respectively the downstream and upstream endpoints of the platoon. At the platoon location, the road capacity is reduced proportionally to the number of lanes occupied by the platoon, and the platoon acts as a flux constraint on the interval $[z_U(t), z_D(t)]$. The resulting coupled PDE-ODE model reads

$$\partial_t u + \partial_x F(t, x, u) = 0, \qquad\qquad (t, x) \in \mathbb{R}^+ \times \mathbb{R}, \qquad\qquad (5.14\text{a})$$

$$u(0, x) = u_0(x), \qquad\qquad x \in \mathbb{R}, \qquad\qquad (5.14\text{b})$$

$$\dot{z}_U(t) = v_U(t, u(t, z_U(t)+)), \qquad\qquad t \in \mathbb{R}^+, \qquad\qquad (5.14\text{c})$$

$$z_U(0) = z_U^0, \qquad\qquad (5.14\text{d})$$

$$\dot{z}_D(t) = v_D(t, u(t, z_D(t)+)), \qquad\qquad t \in \mathbb{R}^+, \qquad\qquad (5.14\text{e})$$

$$z_D(0) = z_D^0. \qquad\qquad (5.14\text{f})$$

Above, the space-time discontinuous flux function F is defined as

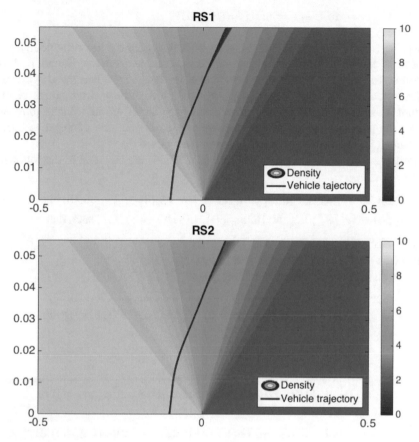

Fig. 5.8 Spatio-temporal evolution of traffic density and moving bottleneck trajectory given by $\mathcal{R}^{\omega}_{ARZ1}$ (top) and $\mathcal{R}^{\omega}_{ARZ2}$ (bottom) corresponding to the data $(u_L, v_L) = (9, 1)$ for $x < 0$, $(u_R, v_R) = (2, 8)$ for $x > 0$, $\omega = 4$, $\alpha = 0.5$, $u_{\max} = 15$ and $y_0 = -0.1$

$$F(t, x, u) := \begin{cases} f(u) & \text{if } x \notin [z_U(t), z_D(t)], \\ f_\alpha(u) := \alpha f(u/\alpha) & \text{if } x \in [z_U(t), z_D(t)]. \end{cases} \tag{5.15}$$

To comply with the varying road capacity, we have to consider initial data u_0 such that

$$\begin{aligned} u_0(x) &\in [0, \alpha u_{\max}] \text{ if } x \in [z_U^0, z_D^0], \\ u_0(x) &\in [0, u_{\max}] \quad \text{otherwise.} \end{aligned} \tag{5.16}$$

The dynamics of the platoon ending points is governed by (5.14c)–(5.14e), where

$$v_U(t, u) := \max \left\{ \omega_U(t), -f_\alpha(u)/(u_{\max} - u) \right\}, \tag{5.17}$$

$$v_D(t, u) := \min\{\omega_D(t), v(u)\},\tag{5.18}$$

where $\omega_U(t) \in [-v_{max}, v_{max}]$ and $\omega_D(t) \in [0, v_{max}]$ are the controllable maximal speeds of the upstream and downstream endpoints respectively. Equation (5.18) accounts for the fact that the platoon cannot move quicker than the downstream traffic velocity. Moreover, the speed ω_D is constrained to be positive, since vehicles cannot move backwards. On the other hand, if vehicles are allowed to join (and leave) the platoon, ω_U may take negative values. In the case of negative speed, condition (5.17) ensures that the problem is well-posed.

Following [243], weak entropy solutions of (5.14) are intended in the following sense:

Definition 26 A weak entropy solution to (5.14)–(5.15)–(5.16) is a triple $(u, z_U, z_D) \in \mathbf{C}^0\left(\mathbb{R}^+; \mathbf{L}^1_{loc}(\mathbb{R}; [0, u_{max}])\right) \times \left(\mathbf{W}^{1,\infty}(\mathbb{R}^+; \mathbb{R})\right)^2$ such that

(i) $u \in \mathbf{L}^\infty\left(\mathbb{R}^+; \mathbf{BV}(\mathbb{R}; [0, u_{max}])\right)$;
(ii) $u(t, x) \in [0, \alpha u_{max}]$ for a.e. $x \in [z_U(t), z_D(t)]$ and $t > 0$;
(iii) for all $\kappa \in \mathbb{R}$ and all test functions $\varphi \in \mathbf{C}^1_c(\mathbb{R}^2; \mathbb{R}^+)$ it holds

$$\int_{\mathbb{R}^+} \int_{\mathbb{R}} \left(|u - \kappa|\, \partial_t \varphi + \mathrm{sgn}(u - \kappa)\left(F(t, x, u) - F(t, x, \kappa)\right) \partial_x \varphi\right)\, \mathrm{d}x\, \mathrm{d}t$$

$$+ \int_{\mathbb{R}} |u_0 - \kappa|\, \varphi(0, x)\, \mathrm{d}x$$

$$+ \int_{\mathbb{R}^+} |F(t, z_U(t)+, \kappa) - F(t, z_U(t)-, \kappa)|\, \varphi(t, z_U(t))\, \mathrm{d}t$$

$$+ \int_{\mathbb{R}^+} |F(t, z_D(t)+, \kappa) - F(t, z_D(t)-, \kappa)|\, \varphi(t, z_D(t))\, \mathrm{d}t \geq 0;$$

(iv) z_U and z_D are Carathéodory solutions of (5.14c)–(5.14d), respectively (5.14e)–(5.14f), i.e., for a.e. $t \in \mathbb{R}^+$ it holds

$$z_U(t) = z_U^0 + \int_0^t v_U(s, u(s, z_U(s)+))\, \mathrm{d}s,$$

$$z_D(t) = z_D^0 + \int_0^t v_D(s, u(s, z_D(s)+))\, \mathrm{d}s.$$

The construction of the corresponding Riemann solvers is detailed in [243, Section 3]. Figure 5.9 provides an illustration of the entropy weak solution for $f(u) = u(1 - u)$ and $\alpha = 0.5$ corresponding to the following initial data:

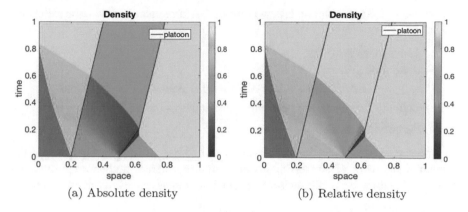

(a) Absolute density   (b) Relative density

Fig. 5.9 Space-time evolution of the solution to (5.14) corresponding to the initial datum (5.19): plot (**a**) displays the absolute density values $u(t, x)$ everywhere, plot (**b**) accounts for the relative density $u(t, x)/\alpha u_{\max}$ at the platoon location, accounting for the reduced road capacity

$$u_0(x) = \begin{cases} 0.3 & \text{if } x < 0.2, \\ 0.4 & \text{if } 0.2 \leq x < 0.5, \\ 0.5 & \text{if } 0.5 \leq x < 0.8, \\ 0.9 & \text{if } x \geq 0.8, \end{cases} \qquad z_U^0 = 0.2, \qquad z_D^0 = 0.5, \qquad (5.19)$$

and control values $\omega_D = 0.3$ and $\omega_U = 0.1$. At the back-end of the platoon a congestion appears. For the downstream endpoint of the platoon, we have a rarefaction wave followed by a shock. When the platoon reaches the downstream congestion, the front-end point of the platoon slows down adapting its speed to the downstream traffic. Since the speeds of the initial and final points of the platoon are different, the length of the platoon changes during the simulation. In particular, the speed of the upstream endpoint is not affected by the surrounding traffic conditions.

5.3 Numerical Methods for Moving Bottlenecks

In this section we provide numerical schemes for the moving bottleneck models introduced in Sect. 5.2. Simulations are performed using algorithms based on the Godunov scheme [154]. All the methods for the numerical approximation of the moving bottleneck models presented in this section are divided into two steps:

1. **Numerical solution of the conservation law**: using an *ad-hoc* finite volume method, for example, the Godunov scheme.
2. **Numerical solution of the ODE**: typically solved by an Euler-type method based on the approximate PDE solution computed at the previous step.

To discretize the conservation law, we introduce a numerical grid with the following notation:

- Δx is the space grid size;
- Δt is the time grid size;
- $(t^n, x_j) = \left(n\Delta t, \left(j + \frac{1}{2}\right)\Delta x\right)$ are the grid points for $n \in \mathbb{N}$ and $j \in \mathbb{Z}$, respectively the number of time and space nodes of the grid, while $x_{j-\frac{1}{2}} = j\Delta x$ are the cell interfaces.

In order to approximate the conservation law (5.3a), we place ourselves in the general framework of conservative finite volume schemes and first approximate the initial datum by a piecewise constant function given by its average on the discretization cells $C_j = \left[x_{j-\frac{1}{2}}, x_{j+\frac{1}{2}}\right]$, namely

$$u_j^0 = \frac{1}{\Delta x} \int_{x_{j-\frac{1}{2}}}^{x_{j+\frac{1}{2}}} u_0(x)\, dx, \quad j \in \mathbb{Z}.$$

To compute the approximation u_j^n of the average value of the exact solution u at time t^n on the cell C_j for $j \in \mathbb{Z}$ and $n \geq 1$, we then integrate the conservation law. Using Green's formula, we naturally end up with an iterative procedure of the form

$$u_j^{n+1} = u_j^n - \frac{\Delta t}{\Delta x}\left(F_{j+\frac{1}{2}}^n - F_{j-\frac{1}{2}}^n\right), \tag{5.20}$$

where the numerical fluxes $F_{j+\frac{1}{2}}^n$ represent an approximate value of the exact flux that passes through the interface $x_{j+\frac{1}{2}}$ in the time interval $[t^n, t^{n+1}[$.

5.3.1   A Coupled Godunov-ODE Scheme for Model (5.1)

We first describe the Godunov numerical scheme for the space dependent flux function (5.1a). We consider a cell-centered discretization of the flux function [24] and we proceed as follows. For each point $x \in \mathbb{R}$, we define the corresponding demand and supply functions as

$$D_\varphi(x, y, u) = \begin{cases} f_\varphi(x, y, u) & \text{if } u < u_{cr}, \\ f_\varphi(x, y, u_{cr}) & \text{if } u \geq u_{cr}, \end{cases}$$

$$S_\varphi(x, y, u) = \begin{cases} f_\varphi(x, y, u_{cr}) & \text{if } u < u_{cr}, \\ f_\varphi(x, y, u) & \text{if } u \geq u_{cr}, \end{cases}$$

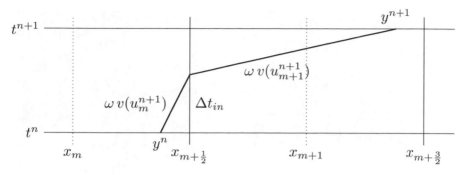

Fig. 5.10 Construction of the approximate bottleneck trajectory y^n

where $u_{cr} \in [0, u_{max}]$ is the point of maximum of $f(u)$. Then the Godunov numerical flux at the cell interface $j + \frac{1}{2}$ and time t^n can be computed as

$$F^n_{j+\frac{1}{2}} = \min \left\{ D_\varphi(x_j, y^n, u^n_j), S_\varphi(x_{j+1}, y^n, u^n_{j+1}) \right\},$$

where y^n denotes the approximated bottleneck position at time t^n. The density can then be updated using (5.20), under the strong CFL condition $\Delta t \leq 0.5 \, \Delta x / v_{max}$.

To update the bottleneck position, let $m \in \mathbb{Z}$ be such that $y^n \in C_m$. Using the updated density value u^{n+1}_m, we move the bottleneck position at speed $\omega \, v(u^{n+1}_m)$ until it reaches the cell interface $m + \frac{1}{2}$ at $t^n + \Delta t_{in}$ given by

$$\Delta t_{in} = \frac{x_{m+\frac{1}{2}} - y^n}{\omega \, v(u^{n+1}_m)},$$

after which it continues at speed $\omega \, v(u^{n+1}_{m+1})$, see Fig. 5.10. Therefore we set

$$y^{n+1} = y^n + \min\{\Delta t, \Delta t_{in}\} \, \omega \, v(u^{n+1}_m) + \max\{\Delta t - \Delta t_{in}, 0\} \, \omega \, v(u^{n+1}_{m+1}).$$

Figure 5.11 shows a controlled vehicle interacting with a shock and a rarefaction wave.

5.3.2 A Conservative Scheme for Non-Classical Solutions to the PDE-ODE Models with Flux Constraint

The PDE-ODE models (5.3), (5.8) and (5.14) can be numerically solved using the following approach, here detailed for the scalar case.

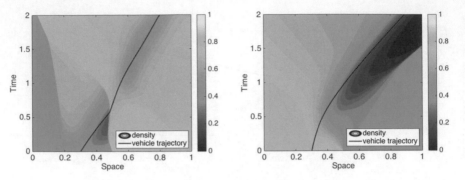

Fig. 5.11 Model (5.1): $x - t$ plots for the interaction of the controlled vehicle with a shock (left) and a rarefaction (right), with $v(u) = 1 - u$, $\kappa = 0.6$ and $\delta = 0.25$

Numerical Scheme for the PDE (5.3a) We use the classical Godunov scheme [154] away from the non-classical shocks, i.e. away from the bottleneck position. In the vicinity of the controlled vehicle position, where non-classical shock waves may arise, we refer to the approach introduced in [45], which consists in adding details in the piecewise constant representation (on each cell C_j) of the approximate solution. The discontinuities are reconstructed in the relevant cells C_j and used to define the numerical fluxes $F^n_{j+\frac{1}{2}}$ instead of simply using the constant values u^n_j and u^n_{j+1}. This allows to exactly capture isolated non-classical discontinuities.

Assume that at time t^n, the bottlenck position y^n is located in the cell C_m for some $m \in \mathbb{Z}$. According to the Riemann solver provided in Sect. 5.2.2, Definition 23, a non-classical shock could appear locally around the bus whenever the condition

$$f(u^n_m) > F_\alpha(\omega(t^n)) + \omega(t^n)u^n_m, \tag{5.21}$$

is satisfied. The idea is to consider u^n_m not only as an approximation of the average value of the solution at time t^n on the cell C_j, but also as resulting from information on the structure of the exact Riemann solution $\mathcal{R}^{\omega(t^n)}(u^n_{m-1}, u^n_{m+1})$ associated with the states u^n_{m-1} and u^n_{m+1}. Therefore, if also

$$f(\mathcal{R}(u^n_{m-1}, u^n_{m+1})(\omega(t^n))) > F_\alpha(\omega(t^n)) + \omega(t^n)\mathcal{R}(u^n_{m-1}, u^n_{m+1})(\omega(t^n)), \tag{5.22}$$

holds, in the cell C_m we replace u^n_m by the left and right states $u^n_{m,l} = \hat{u}_{\omega(t^n)}$ and $u^n_{m,r} = \check{u}_{\omega(t^n)}$ corresponding to the non-classical shock appearing in the constrained Riemann solver associated with u^n_{m-1} and u^n_{m+1}. To guarantee mass conservation, the reconstructed discontinuity must be located inside C_m at the position

$$\bar{x}_m = x_{m-\frac{1}{2}} + d^n_m \Delta x, \tag{5.23}$$

Fig. 5.12 Reconstruction of
a non-classical shock

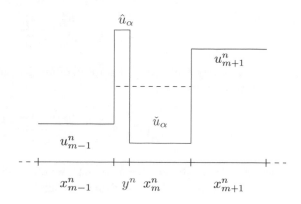

where $d^m \in [0, 1]$ is defined by

$$d_m^n u_{m,l}^n + (1 - d_m^n)u_{m,r}^n = u_m^n, \tag{5.24}$$

see Fig. 5.12.

We need now to define the numerical flux at the cell interfaces $x_{m\pm\frac{1}{2}}$. Let us first recall that Godunov's flux function (or any monotone and consistent numerical flux) is used to define $F_{j+1/2}^n$, $j \neq m$, under the classical CFL condition

$$\Delta t \max_{j \in \mathbb{Z}} \left| f'(u_j^n) \right| \leq \frac{1}{2} \Delta x, \tag{5.25}$$

and we keep the same definition for $F_{m+1/2}^n$ whenever the conditions (5.21) and (5.22) are not both satisfied simultaneously. Let us denote by

$$\Delta t_{m+\frac{1}{2}} = \frac{1 - d_m^n}{V_b} \Delta x$$

the time needed by the reconstructed discontinuity to reach the interface. The numerical flux at $x_{m+\frac{1}{2}}$ is then given by

$$\Delta t\, F_{m+\frac{1}{2}}^n = \min\left(\Delta t_{m+\frac{1}{2}}, \Delta t\right) f(u_{m,r}^n) + \max\left(\Delta t - \Delta t_{m+\frac{1}{2}}, 0\right) f(u_{m,l}^n). \tag{5.26}$$

Besides, we set $F_{m-\frac{1}{2}}^n = f(u_{m,l}^n)$.

To overcome problems induced by numerical diffusion around classical shocks, the reconstruction technique can also be applied to classical shocks. Given a cell C_j for some $j \in \mathbb{Z}$, $j \neq m$, such that $u_{j-1}^n < u_{j+1}^n$, we introduce the left and right traces $u_{j,l}^n = u_{j-1}^n$ and $u_{j,r}^n = u_{j+1}^n$ of a reconstructed discontinuity and we define d_j^n by

$$d_j^n = \frac{u_{j,r}^n - u_j^n}{u_{j,r}^n - u_{j,l}^n}. \tag{5.27}$$

Let us denote by $\lambda(u_{j,l}^n, u_{j,r}^n)$ the speed of the discontinuity given by the Rankine–Hugoniot condition, i.e.,

$$\lambda(u_{j,l}^n, u_{j,r}^n) := \frac{f(u_{j,l}^n) - f(u_{j,r}^n)}{u_{j,l}^n - u_{j,r}^n}.$$

Then, if it is actually possible to reconstruct such a discontinuity within the cell C_j, that is to say if $d_j^n \in [0, 1]$, the numerical fluxes at $x_{j \pm \frac{1}{2}}$ are defined by

- if $\lambda(u_{j,l}^n, u_{j,r}^n) \geq 0$,

$$\Delta t \, F_{j+\frac{1}{2}}^n = \min(\Delta t_{j+\frac{1}{2}}, \Delta t) f(u_{j,r}^n) + \max(\Delta t - \Delta t_{j+\frac{1}{2}}, 0) f(u_{j,l}^n), \tag{5.28}$$

with $\Delta t_{j+\frac{1}{2}} = \dfrac{1 - d_j^n}{\lambda(u_{j,l}^n, u_{j,r}^n)} \Delta x$.

- if $\lambda(u_{j,l}^n, u_{j,r}^n) \leq 0$,

$$\Delta t \, F_{j-\frac{1}{2}}^n = \min(\Delta t_{j-\frac{1}{2}}, \Delta t) f(u_{j,l}^n) + \max(\Delta t - \Delta t_{j-\frac{1}{2}}, 0) f(u_{j,r}^n), \tag{5.29}$$

with $\Delta t_{j-\frac{1}{2}} = \dfrac{d_j^n}{-\lambda(u_{j,l}^n, u_{j,r}^n)} \Delta x$,

where, with some abuse of notation, we mean that if $\lambda(u_{j,l}^n, u_{j,r}^n) = 0$ then $F_{j+\frac{1}{2}}^n = f(u_{j,r}^n)$ and $F_{j-\frac{1}{2}}^n = f(u_{j,l}^n)$.

Numerical Scheme for the ODE (5.1b) To precisely track the controlled vehicle at each time step, we update its position y^n by studying interactions between the vehicle's trajectory and the density waves within the corresponding cell. We distinguish two cases:

1. Inequality (5.21) is satisfied. Then the vehicle moves at velocity $\omega(t^n)$ and we update its position as $y^{n+1} = \omega(t^n) \Delta t^n + y^n$.
2. Condition (5.21) is not satisfied. In this case the PDE solution is classical. We have to distinguish two situations: either $y^n \in [x_{m-\frac{1}{2}}, x_m^n[$ or $y^n \in [x_m^n, x_{m+\frac{1}{2}}[$. If no interaction occurs between the wave originating at the corresponding cell interface and the controlled vehicle in Δt^n, the ODE is solved by

$$y^{n+1} = y^n + \min\{\omega(t^n), v(u_m^n)\} \Delta t.$$

Otherwise, we check if the wave is a shock or a rarefaction:

- If the wave is a shock, we compute the incremental interaction time $\bar{t}$ and the vehicle's trajectory is given by

$$y^{n+1} = y^n + \min\{\omega(t^n), v(u_m^n)\}\bar{t} + \min\{\omega(t^n), v(u_{m\pm1}^n)\}(\Delta t - \bar{t}).$$

- If the wave is a rarefaction, first of all, we observe that if this wave originated at $x_{m-\frac{1}{2}}$ and an interaction occurs, the controlled vehicle must travel at its maximal velocity $\omega(t^n)$, and it will keep this velocity, as in Case 1. Therefore, we focus on the case of a rarefaction originating at $x_{m+\frac{1}{2}}$. If the vehicle is initially traveling at speed $\omega(t^n)$, it will keep this velocity after the interaction, see Case 1. When the vehicle does not travel at the constant speed $\omega(t^n)$, we compute its first and last interaction points with the rarefaction wave, respectively $(\bar{t}, \bar{x})$ and $(\bar{\bar{t}}, \bar{\bar{x}})$, to evaluate the exact trajectory. Then:

  - If $\bar{\bar{t}} \geq \Delta t$, the new position is given by $y^{n+1} = \tilde{y}(\Delta t)$ with $\tilde{y}(\Delta t)$ given by setting $t = \Delta t$ in

$$\tilde{y}(t) = x_{m+\frac{1}{2}} + (t - t^n) - \sqrt{t - t^n}\left(\frac{\bar{t} - t^n + x_{m+\frac{1}{2}} - \bar{x}}{\sqrt{\bar{t} - t^n}}\right).$$

  - If $\bar{\bar{t}} < \Delta t$, then $y^{n+1} = \tilde{y}(\bar{\bar{t}}) + (\Delta t - \bar{\bar{t}})\min\{\omega(t^n), v(u_{m+1}^n)\}$.

The cell index m is then updated according to the new position of the bottleneck. In Fig. 5.13, we can see the interaction of the bus with a shock and with a rarefaction.

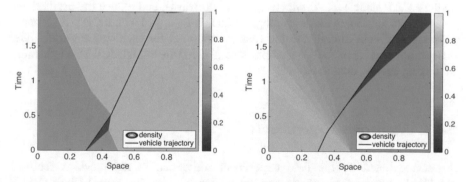

Fig. 5.13 Model (5.3): $x - t$ plots for the interaction of the controlled vehicle with a shock (left) and a rarefaction (right), with $v(u) = 1 - u$ and $\alpha = 0.6$

5.4   Traffic Management by Controlled Vehicles

5.4.1   Field Experiments

Stop and go waves are instabilities of traffic that travel backwards along the road [190, 268]. Often called phantom jams, this phenomenon emerges from the collective dynamics of the drivers on the road. In [262], a seminal field experiment proved that is possible to use an automated and controlled vehicle to dissipate them. A series of experiments were conducted in a ring road setting to show that a *Cognitive and Automated Test* (CAT) vehicle (see Fig. 5.14), properly controlled is able to dampen stop and go waves. The experiment setup follows the work done by [263]: 22 vehicles are placed in a ring road formation on single-lane circular track of 260 meters, see Fig. 5.15. Each vehicle is equipped with OBD-II data loggers that collect GPS locations and fuel consumption for each one of the vehicles. In the center of the track, a 360 degree camera records each experiment. After post-processing, the videos will be used to generate vehicle trajectories. One of the vehicles is the University of Arizona self-driving capable CAT vehicle.

At the beginning of each experiment, the drivers are told to drive as they normally would. The goal of the experiments is to create stop and go waves and then dissipate them via control of the CAT vehicle. At the beginning of the experiment, the vehicles are equidistant and at rest. In Fig. 5.16, one can see the trajectories of the vehicles during an experiment. After the first minute, we observe the creation of the stop and go wave which becomes more evident if one observes the corresponding velocity oscillations in Fig. 5.17. In particular, after the wave starts, the velocity oscillates between 0 and 11 m/s. At that point, the CAT vehicle controller is activated and the wave dissipates. After 350 seconds, the control is disabled and the wave reappears again until the end of the experiment. From Figs. 5.16 and 5.17, we can see how a single CAT vehicle can dissipate stop and go waves, and reduce the oscillations of the velocities. Given that each single vehicle is equipped with a OBD-II scanner, we were also able to compute the fuel consumption for each single vehicle. We observe a reduction of fuel consumption of 43% for the total fleet, with a reduction of braking events equal to 98%.

5.4.2   Numerical Experiments

As demonstrated by the field experiments described in the previous section, see [262], a controlled vehicle can act as moving Variable Speed Limit (VSL) to optimize the traffic flow on the overall section of the highway taken into consideration, by reducing stop and go waves, congestion and pollution. Referring to (5.3), here the control variable is the maximal speed ω of the moving bottleneck in (5.1b).

Fig. 5.14 Arizona CAT vehicle *Media credit: John de Dios, Alan Davis*

Fig. 5.15 Car alignment on a single-lane ring road track *Media credit: John de Dios, Alan Davis*

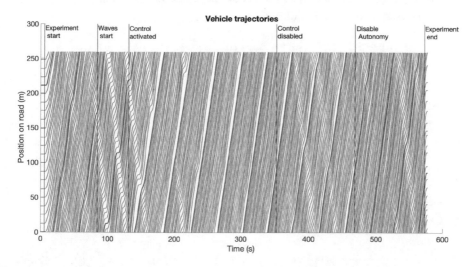

Fig. 5.16 Trajectories of each single vehicle on the ring road, the CAT vehicle is in blue, while the human driven vehicles trajectories are in gray

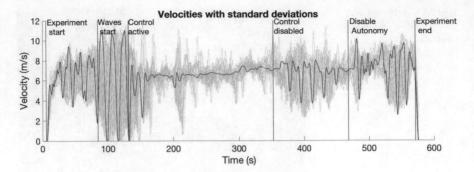

Fig. 5.17 Speed profiles for each single vehicle on the ring road, the CAT vehicle profile is in blue, while the human driven vehicles are in gray

As an example, let us consider a stretch of road corresponding to the space interval $[a, b]$ and a time horizon T_f. To evaluate the impact of the control policy, we focus on the following traffic performance indexes:

- *Total Fuel Consumption (TFC)*. We rely on a model that expresses the fuel consumption as a function of the speed [246] and we define

$$TFC(\omega) := \int_0^{T_f} \int_a^b u(t, x) K(v(u(t, x))) \, \mathrm{d}x \, \mathrm{d}t, \tag{5.30}$$

where the fuel consumption depends on the traffic mean velocity according to the following polynomial expression

$$K(v) = 5.7 \cdot 10^{-12} \cdot v^6 - 3.6 \cdot 10^{-9} \cdot v^5 + 7.6 \cdot 10^{-7} \cdot v^4$$
$$- 6.1 \cdot 10^{-5} \cdot v^3 + 1.9 \cdot 10^{-3} \cdot v^2 + 1.6 \cdot 10^{-2} \cdot v + 0.99,$$

see Fig. 5.18.

- *Average Travel Time (ATT)*. This is computed as

$$ATT(\omega) := \int_0^{T_f} \int_a^b \frac{1}{v(u(t, x))} \, \mathrm{d}x \, \mathrm{d}t, \tag{5.31}$$

see, for example, [97].

- *Queue length*. This is expressed by

$$\Psi(\omega) = \frac{1}{T_f} \int_0^{T_f} \int_a^b \phi(u(t, x)) \, \mathrm{d}x \, \mathrm{d}t, \tag{5.32}$$

where

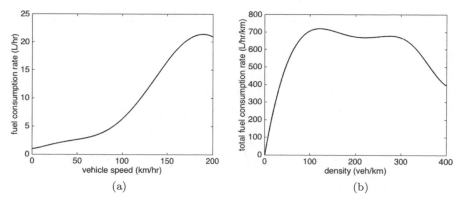

Fig. 5.18 Plots of the fuel consumption rate functions. (**a**) Fuel consumption rate $K(v)$. (**b**) Total FC rate $uK(v(u)))$

$$\phi(r) = \begin{cases} 0 & r < u_{out} - \delta, \\ \frac{1}{\delta}(r - u_{out} + \delta) & u_{out} - \delta \leq r \leq u_{out}, \\ 1 & u_{out} < r \leq u_{max}, \end{cases} \tag{5.33}$$

for some $\delta > 0$ small (see [83]). This functional measures the length of the traffic jam arising upward a fixed bottleneck located at $x = b$, which is expressed by the boundary constraint $f(u(t, b-)) \leq F_{out}$, for some given $F_{out} \leq f_{max} := \max_{u \in [0, u_{max}]} f(u)$. (For the definition and treatment of the constrained Initial-Boundary value problem see [78].) In (5.33), the u_{out} corresponds to the congestion traffic density, which satisfies $f(u_{out}) = F_{out}$, $u_{out} \geq u_{cr}$.

To evaluate the dependence of the above cost functionals on the control variable ω, we consider a section of highway of length $L = 50 \, km$ with three lanes (therefore $\alpha = 2/3$), where we set $v(u) = v_{max}(1 - u/u_{max})$ with $v_{max} = 140$ km/hr and $u_{max} = 400 \, vehicles/km$. The initial traffic conditions are set to $u_0 = 0.3 \, u_{max}$, and the controlled vehicle is placed at $y_0 = 2 \, km$ from the beginning of the road section. Boundary conditions are prescribed in terms of boundary fluxes f_{in} and f_{out}, setting

$$f_{in}(t) = \begin{cases} f^{max} & \text{if } t \leq 0.5 \, T_f, \\ 0 & \text{if } t > 0.5 \, T_f, \end{cases}$$

and $f_{out}(t) = 0.5 \, f^{max}$ for $t \in [0, T_f]$, where the time horizon $T_f = 1$ hr.

Figure 5.19 shows the dependence of the functionals (5.30)–(5.32) on the controlled vehicle maximal speed $\omega \in [0, 140]$.

In order to get benefits in terms of some prescribed cost function, in this case the total fuel consumption (5.30), we design a *Model Predictive Control* (MPC) strategy, which is quite common in traffic management (see, among others, [31, 47, 131]).

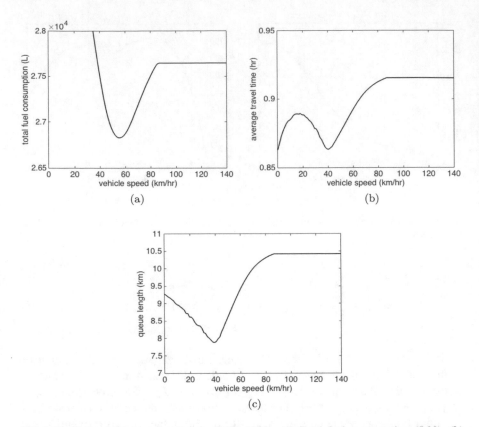

Fig. 5.19 Plots of the cost functionals (5.30)–(5.32). (**a**) Total fuel consumption (5.30). (**b**) Average travel time (5.31). (**c**) Queue length (5.32)

To this end, system (5.3) is discretized using the conservative scheme presented in Sect. 5.3.2. The optimization algorithm is run on a prediction horizon $\Delta T = 15$ min and the optimal control value is applied for the time interval $\Delta \tau = 5$ min before the state is updated and the optimal control re-evaluated. See [242] for further details. Figure 5.20 shows the comparison of traffic evolutions in the case of fixed vehicle speed $\omega(t) = 80$ km/hr for all $t \in [0, T_f]$ (left) and in the optimal controlled case in which the vehicle speed is computed by MPC (right). In particular, we can notice that the congested region is reduced by the implementation of the control strategy. Indeed, we observe that the enforcement of the optimal control allows not only to improve the optimized functional (in this case the TFC), but also to improve the other metrics considered, see Table 5.1.

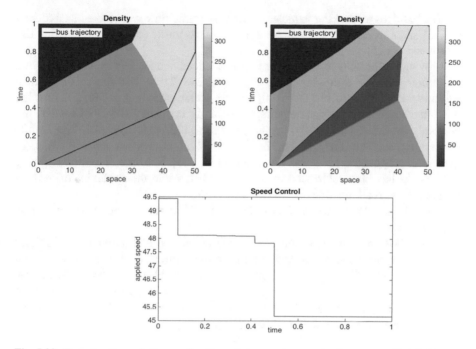

Fig. 5.20 Top: density evolution as function of time and space in the uncontrolled (left) and controlled (right) cases. Bottom: applied maximum vehicle speed resulting from the MPC procedure

Table 5.1 Comparison between cost functionals in the controlled and uncontrolled case

| | ATT | Ψ | TFC | TFC reduction |
	[hr]	[km]	[liters]	%
Uncontrolled	0.9107	10.18	$2.7413 \cdot 10^4$	0
Controlled	0.8579	7.66	$2.6852 \cdot 10^4$	2.05

5.5 Bibliographical Notes

Conservation laws with point flux constraint at a fixed space location were first introduced in [78], where a well-posedness result for scalar solutions corresponding to **BV** initial data and flux constraint was based on the wave-front tracking method and the doubling of variables technique. This result was then extended to an L^∞ framework in [17], were adapted finite volume schemes are proposed. The scalar setting was then further extended in [15, 68] for applications to pedestrian flow modeling, see also [14, 104] for applications to other settings. Regarding systems, [16, 105, 124, 137, 145], among others, deal with the ARZ system and its extensions.

The problem of tracking vehicles in the bulk traffic flow was studied analytically in [82] and numerically in [58]. An application to traffic states reconstruction can be found in [115].

Moving bottlenecks models were first introduced and studied in the engineering literature, see, for example, [101, 102, 203, 208–210]. The first mathematical settings were proposed in [112, 200] for the scalar case, and in [42, 272] for the ARZ model. Well-posedness results for strongly coupled PDE-ODE models are currently available only in the scalar case, see [112, 114, 138, 222, 223], while numerical treatments are proposed in [67, 111, 272]. Numerical approximations of the model proposed in [200] can be found in [56, 146]. The case of multiple moving bottlenecks has been addressed in [56, 113, 146], while the dynamics of moving bottlenecks at road junctions is analyzed by [125].
A similar PDE-ODE model is proposed in [201, 202] to model bounded acceleration in traffic flow.

Traffic flow management and control via connected and autonomous vehicles is currently a hot topic in transport engineering research. A number of results are available via stability analysis or simulation, such as [109, 163, 264, 274]. Some control approaches are based on implementing variable speed limit strategies, such as [31, 168, 273], or via jam absorption [172, 235]. Finally, experiments showed the effectiveness of the approach, see [261, 262, 275].

Chapter 6
Control Problems for Hamilton-Jacobi Equations *Co-authored by Alexander Keimer*

6.1 Introduction

In this chapter, we introduce Hamilton-Jacobi PDEs. These PDEs are related to conservation laws and their solutions are the anti-derivative (in space) of the Entropy solutions of the corresponding conservation law, given that some assumptions are satisfied.

Roughly, a Hamilton-Jacobi PDE reads for the Cauchy problem on $\mathbb{R}$ for a given flux function $f : \mathbb{R} \to \mathbb{R}$ as

$$\partial_t q(t, x) + f\big(\partial_x q(t, x)\big) = 0 \qquad (t, x) \in (0, T) \times \mathbb{R}$$
$$q(0, x) = q_0(x) \qquad x \in \mathbb{R}.$$

Assuming that the solutions are significantly smooth, we can differentiate the PDE with respect to the spatial variable and obtain

$$\partial_t \, \partial_x \, q(t, x) + \partial_x \, f\big(\partial_x \, q(t, x)\big) = 0 \qquad (t, x) \in (0, T) \times \mathbb{R}$$
$$\partial_x \, q(0, x) = q_0'(x) \qquad x \in \mathbb{R}.$$

Setting $\partial_x q \equiv u$ for a given function $u : (0, T) \times \mathbb{R} \to \mathbb{R}$ one ends up with

$$\partial_t u(t, x) + \partial_x f(u(t, x)) = 0 \qquad (t, x) \in (0, T) \times \mathbb{R}$$
$$u(0, x) = q_0'(x) \qquad x \in \mathbb{R},$$

the conservation law in u with flux function f and initial datum $q_0'(x), x \in \mathbb{R}$. Having a closer look into the theory of Hamilton-Jacobi PDEs, one major advantage is that the solutions to these equations remain Lipschitz continuous in case the initial

© The Author(s) 2022
A. Bayen et al., *Control Problems for Conservation Laws with Traffic Applications*,
PNLDE Subseries in Control 99, https://doi.org/10.1007/978-3-030-93015-8_6

datum is Lipschitz continuous (and in the case one is in the bounded domain case, also the boundary datum).

Even more, there is an explicit solution formula taking advantage of the Legendre Fenchel dual to write down the solution at every given time-space point as an convex optimization problem. This is of great importance as it not only allows to evaluate solutions without a predetermined numerical grid, the specific solution structure enables it also to study details and behavior of solutions.

These loosely described approaches will be made rigorous in the next section where we introduce the concept of strong and generalized solutions.

6.2 Strong Solutions

We start with the problem setup and some assumptions on the flux function which contain convexity. The same results also hold for concave functions under minor modifications.

Definition 27 (The Hamilton-Jacobi PDE with Initial Datum) Given a flux function $f : \mathbb{R} \to \mathbb{R}$ and initial datum $q_0 : \mathbb{R} \to \mathbb{R}$ we call the following initial value problem

$$\partial_t q(t, x) + f\big(\partial_x q(t, x)\big) = 0 \qquad\qquad (t, x) \in (0, T) \times \mathbb{R}$$

$$q(0, x) = q_0(x) \qquad\qquad x \in \mathbb{R}.$$

a Hamilton-Jacobi PDE, and $q : (0, T) \times \mathbb{R} \to \mathbb{R}$ its solution.

To guarantee the existence and uniqueness of solutions, we require some assumptions on initial datum and flux function. This is detailed in the following

Assumption 1 (Flux Function and Initial Datum) *We assume that*

1. $q_0 \in W_{loc}^{1,\infty}(\mathbb{R}) : q_0' \in L^\infty(\mathbb{R})$ *2.* $f \in W_{loc}^{1,\infty}(\mathbb{R})$ *strictly convex satisfying* $\lim_{x \to \pm\infty} \frac{f(x)}{|x|} = \infty$.

For the explicit solution formula for every fixed space-time point we require the convex conjugate of the function which we define as follows.

Definition 28 (Convex Conjugate or the Legendre Fenchel Transform) Suppose that f satisfies item 2 in assumption 1, i.e., is in particular strictly convex and let $I \subseteq \mathbb{R}$ be a closed interval. Then, we define the Legendre Fenchel transform f^* of f on I as

$$f^*(x) := \sup_{u \in I} \{ux - f(u)\}, \quad x \in \mathrm{Dom}(f^*), \tag{6.1}$$

where the domain for the Legendre transform f^* on I is defined as

$$\text{Dom}(f^*) := \left\{ x^* \in \mathbb{R} : \sup_{x \in I} \left\{ x x^* - f(x) \right\} < \infty \right\}. \tag{6.2}$$

Remark 13 (Convex Conjugate) Note that due to item 2, $\text{Dom}(f^*)$ is an interval in $\mathbb{R}$ or entire $\mathbb{R}$. Furthermore, f^* is also a convex function.

Given these definitions and assumptions we can state the main Theorem 33:

Theorem 33 (Existence/Uniqueness of Strong Solutions and the Lax-Hopf Formula) *Let Assumption 1 hold. Then, the initial value problem in Definition 27 admits a unique strong solution* $q \in W_{loc}^{1,\infty}((0, T) \times \mathbb{R})$, $\partial_2 q \in L^{\infty}((0, T) \times \mathbb{R})$ *and the solution at time-space point* $(t, x) \in (0, T) \times \mathbb{R}$ *can be posed as*

$$q(t, x) = \min_{y \in \mathbb{R}} q_0(y) + t \cdot f^* \left(\frac{x-y}{t} \right) \tag{6.3}$$

*with f^* as in Definition 28. The latter identity is called **Lax-Hopf** formula.*

Proof The proof can be found in [126, Chapter 3.3].   □

Remark 14 (Interpretations of the Solution Formula and More) As pointed out this solution formula for q requires it to solve a minimization problem at every given point in space-time. However, as f^* is strictly convex and q_0 grows at most linearly due to the assumption on $q_0' \in L^{\infty}(\mathbb{R})$, this minimization problem always possesses one and only one minimal point for every given $(t, x) \in (0, T) \times \mathbb{R}$.

The following theorem makes the connection between the solution of the Hamilton-Jacobi equation and the Entropy solution of the corresponding conservation law. It thus details what had been described in Sect. 6.1. For the general theory of scalar conservation laws and types of solutions, particularly Entropy solutions, we refer the reader to [51].

Theorem 34 (Relation to the Corresponding Conservation Law) *Given Assumption 1, the spatial derivative of the solution stated in Eq. (6.3) in Theorem 33 is the unique Entropy solution of the conservation law*

$$\partial_t u(t, x) + \partial_x f(u(t, x)) = 0 \qquad (t, x) \in (0, T) \times \mathbb{R}$$

$$u(0, x) = q_0'(x) \qquad x \in \mathbb{R}.$$

Proof The proof can be found in many text books, we again refer to [126, Chapter 3.4, Chapter 11].   □

This result is the key why instead of solving a conservation law, one might solve the corresponding Hamilton-Jacobi equation. One advantage of the solution is that it is Lipschitz continuous. Due to the gain in regularity one can apply methods on control and optimization which one could not so easily apply on the level of conservation laws.

In the following, we will also look into the case where boundary datum is prescribed, i.e., when we are on a bounded or semi-bounded domain.

Before addressing the named question we look in detail into the bounded domain case:

6.2.1   The Bounded Domain Case

It is worth mentioning some results for the bounded domain cases:

Theorem 35 (Lax-Hopf Formula on $\mathbb{R}_{>0}$) *Let $T \in \mathbb{R}_{>0}$ and f satisfy Assumption 1. Let the initial value $u_0 \in L^\infty(\mathbb{R}_{>0})$ be given and assume that $u_b \in L^\infty((0, T))$. Then, there exists a unique strong Lipschitz continuous solution to the following Hamilton-Jacobi equation $q : (0, T) \times \mathbb{R}_{\geq 0} \to \mathbb{R}$ with Neumann boundary datum*

$$\partial_t q(t, x) + f(\partial_x q(t, x)) = 0 \qquad\qquad (t, x) \in (0, T) \times \mathbb{R}_{>0}$$

$$q(0, x) = q_0(x) := \int_0^x u_0(y)\mathrm{d}y \qquad\qquad x \in \mathbb{R}_{>0}$$

$$\partial_x q(t, x)\big|_{x=0} = \bar{u}_b(t) \qquad\qquad t \in (0, T).$$

The solution q can be represented by means of a three dimensional restricted minimization problem, for $(t, x) \in \mathbb{R}_{\geq 0} \times [0, T]$

$$q(t, x) = \min \left\{ \min_{y \in \mathbb{R}_{\geq 0}} \left\{ t f^*\left(\tfrac{x-y}{t}\right) + q_0(y) \right\}, \right.$$

$$\left. \min_{\substack{0 \leq t_2 \leq t_1 \leq t \\ a \in \mathbb{R}_{\geq 0}}} \left\{ q_0(a) + f^*\left(\tfrac{-a}{t_2}\right)t_2 + (t - t_1)f^*\left(\tfrac{x}{t-t_1}\right) - \int_{t_2}^{t_1} f(u_b(\theta))\mathrm{d}\theta \right\} \right\}.$$

Furthermore, the spatial derivative of q, i.e.,

$$u(t, x) := \partial_x q(t, x), \qquad (t, x) \in (0, T) \times \mathbb{R}_{>0} \text{ a.e.}$$

is the entropy solution of the conservation law

$$\partial_t u(t, x) + \partial_x f(u(t, x)) = 0 \qquad\qquad (t, x) \in (0, T) \times \mathbb{R}_{>0}$$

$$u(0, x) = u_0(x) = q_0'(x) \qquad\qquad x \in \mathbb{R}_{>0}$$

satisfying the boundary condition in the sense of [29]

$$u(0, t) = \bar{u}_b(t), \quad t \in (0, T) \ almost \ everywhere$$

with

$$\bar{u}_b(t) := \max \left\{ u_b(t), \theta_f \right\}, \quad t \in (0, T) \ a.e., \qquad \theta_f := \text{arg-min}_{x \in \mathbb{R}} f(x)$$

the possibly attained boundary datum.

Proof   The proof can be found in [184].   □

Remark 15 (Some Remarks to the Boundary Datum and the Solution Formula)
The Neumann boundary datum on the level of Hamilton-Jacobi equations can be
interpreted as a Dirichlet boundary datum for the corresponding conservation law.
This is also why the boundary datum can only hold in the sense of [29] and also
needs to be projected into what the flux function f can "handle."

Finally, from the solution formula in Theorem 35 one can see that the solution
is computed in two parts. The first part, i.e., $min_{y \in \mathbb{R}_{\geq 0}} \left\{ t f^* \left(\frac{x-y}{t} \right) + q_0(y) \right\}$ is the
solution which directly propagates the initial datum (compare Eq. (6.3)). The second
part of the minimization is responsible for the interaction of boundary datum and
initial datum.

A further result is available for two sided boundary datum, however, in this case the
solution formula is much more involved as the two boundaries can influence each
other over time. We refer to [185].

6.3   Generalized Solutions

In this section we will introduce more general solutions, the so-called Barron-
Jensen/Frankowska solutions [30, 134] for Hamilton-Jacobi equations. Then, the
spatial derivative might not even exist in a strong sense so that an interpretation on
the level of conservation laws as in the previous section is not possible anymore.
However, from an applied point of view the generalized solutions for the Hamilton-
Jacobi equations have other desirable features. For example, they enable the
reconciliation of data points incompatible with conservation laws, as often measured
in experimental data [74]. In addition, the interpretation of these solutions goes back
to the 1968 seminal article of Karl Moskowitz [234], which gives a practitioner
interpretation to these solutions in terms of vehicle counts (which can be directly
measured with loop detectors). The basic idea follows [21, 72–74] and related work.
As this work is mainly concerned with traffic flow applications, the flux functions
chosen here are concave. In addition, for the conservation law in $q : (0, T) \times X \to \mathbb{R}$
for $t \in [0, T]$ and $x \in X \subset \mathbb{R}$ a bounded interval

$$\partial_t u(t, x) + \partial_x f\big(u(t, x)\big) = 0$$

$$u(0, x) = u_0(x) \tag{6.4}$$

$$+ \text{ boundary conditions}$$

we redefine the corresponding Hamilton-Jacobi equation as

$$\partial_t q(t, x) - f(-\partial_x q(t, x)) = 0 \tag{6.5}$$

so that for smooth solutions we obtain

$$u(t, x) = -\partial_x q(t, x). \tag{6.6}$$

We make this precise requiring the following

Assumption 2 (Assumption on the Flux Function) *Given a maximal density* $u_{\max} \in \mathbb{R}_{>0}$ *we assume that the flux function* f *considered is concave, and Lipschitz, i.e.,* $f \in W^{1,\infty}((0, u_{\max}))$. *As it will be important for the following Legendre Fenchel transform in Definition 29 to have* f *defined on* $\mathbb{R}$ *we extend it by the following procedure. Define*

$$v^{\flat} := \sup_{(x,y)\in[0,u_{\max}]^2} \frac{f(x)-f(y)}{x-y}, \qquad v^{\sharp} := - \inf_{(x,y)\in[0,u_{\max}]^2} \frac{f(x)-f(y)}{x-y} \tag{6.7}$$

we extend f *by* $\tilde{f} : \mathbb{R} \to \mathbb{R}$ *as follows*

$$\tilde{f}(x) = \begin{cases} f(x) & x \in [0, u_{\max}] \\ f(u_{\max}) - v^{\sharp}(x - u_{\max}) & x \in (u_{\max}, \infty) \\ f(0) + v^{\flat}x & x \in (-\infty, 0) \end{cases} \qquad x \in \mathbb{R} \tag{6.8}$$

and have that also $\tilde{f} \in W_{loc}^{1,\infty}(\mathbb{R})$ *is concave. The extension is also illustrated in Fig. 6.1. We also define*

$$f_{\max} := \max_{x\in[0,u_{\max}]} f(x). \tag{6.9}$$

The reformulation of Hamilton-Jacobi equations as in Eq. (6.5) also necessitates the redefinition of the Legendre transform as

Definition 29 (Legendre Fenchel Transform in Traffic Flow Modeling) For a given $u_{\max} \in \mathbb{R}_{>0}$ let a concave flux function $f \in W^{1,\infty}([0, u_{\max}])$ be given and recall its extension $\tilde{f}$ as in Assumption 2, the **Legendre Fenchel** transform reads for $u \in \text{Dom}(f^*)$ as

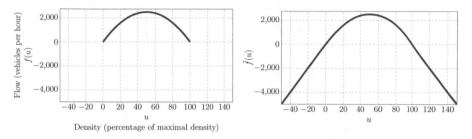

Fig. 6.1 A flux (here Greenshields [159]) and its extension on $\mathbb{R}$ with $u_{\max} = 100$

Fig. 6.2 Triangular flux function with $u_c = 20\%$ and $u_{\max} = 1$

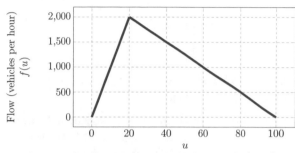

Density (percentage of maximal density)

$$f^*(u) := \sup_{p \in \mathbb{R}} \left\{ p \cdot u + \tilde{f}(p) \right\}, \ \mathrm{Dom}(f^*) := \left\{ u \in \mathbb{R} : \ \sup_{p \in \mathbb{R}} \left\{ p \cdot u + \tilde{f}(p) \right\} < \infty \right\}.$$

For specific flux functions, as, for instance, for triangular flux functions, one can compute the dual function explicitly. We first define a triangular flux function as follows and then compute its dual.

Definition 30 (Triangular Flux Function) We call $f \in W^{1,\infty}([0, u_{\max}])$ for a given $u_{\max} \in \mathbb{R}_{>0}$ a triangular flux function if it satisfies

$$f : \begin{cases} [0, u_{\max}] \ \to \ \mathbb{R} \\ u \ \mapsto \ \begin{cases} vu & u \le u_c \\ w(u - u_{\max}) & \text{else,} \end{cases} \end{cases} \tag{6.10}$$

where the model parameters are the *free flow speed* $v \in \mathbb{R}_{>0}$, the *critical density* $u_c \in [0, u_{\max})$, the *congestion speed* $w \in \mathbb{R}_{<0}$ and the *maximal density* $u_{\max}$, satisfying $vu_c = w(u_c - u_{\max})$. The flux function is illustrated in Fig. 6.2.

To illustrate the complexity of the computations we give in the following an explicit expression for the Legendre Fenchel transform of f:

Remark 16 (Computation of the Dual Function of the Triangular Flux) Let the flux as in Eq. (6.10) be given, and recall the transform in Definition 29 we compute its dual as

$$f^*(u) = \sup_{p \in \mathbb{R}} \left\{ p \cdot u + \begin{cases} vp & p \leq u_c \\ w(p - u_{max}) & p \geq u_c \end{cases} \right\}$$

$$= \sup_{p \in \mathbb{R}} \begin{cases} p(u + v) & p \leq u_c \\ p(u + w) - wu_{max} & p \geq u_c \end{cases}$$

$$= \sup \left\{ \sup_{p \in (-\infty, u_c]} p(u + v), \sup_{p \in [u_c, \infty)} p(u + w) - wu_{max} \right\}$$

$$= \begin{cases} \infty & u < -v \\ u_c(u + v) & -v \leq u \leq -w \\ \infty & u > -w \end{cases},$$

where we have used $vu_c = w(u_c - u_{max})$ so that we obtain as effective domain $\mathrm{Dom}(f^*) = [-v, -w]$.

Assumption 3 (Notation for the Bounded Domain Case) *The spatial interval on which we will consider the previously introduced Eq. (6.4) will be the interval $(A, B) \subset \mathbb{R}$ with $A, B \in \mathbb{R}$, $A < B$. For notational convenience, we will sometimes write*

$$X := (A, B).$$

We then start with a basic definition of value conditions, one key ingredient for the generalized solution:

Definition 31 (Value Condition) For $T \in \mathbb{R}_{>0}$ a value condition $\mathfrak{c} : \mathrm{Dom}(\mathfrak{c}) \subseteq [0, T] \times X \to \mathbb{R}$ is a lower semicontinuous function. We extend $\mathfrak{c}$ by $\tilde{\mathfrak{c}} : [0, T] \times X$ on the entire space-time horizon by defining

$$\tilde{\mathfrak{c}}(t, x) := \begin{cases} \mathfrak{c}(t, x) & \text{for } (t, x) \in \mathrm{Dom}(\mathfrak{c}) \\ \infty & \text{else.} \end{cases} \tag{6.11}$$

For reasons of convenience we replace in the following $\tilde{\mathfrak{c}}$ again by $\mathfrak{c}$ which is now defined on $[0, T] \times X$ but admits as effective domain on the original $\mathrm{Dom}(\mathfrak{c})$.

With this definition we can present the main theorem of this section, the generalized Lax-Hopf formula based on viability theory [22]. This viability epi solution to the Hamilton-Jacobi equation is obtained in epigraphical form from the computation of a viability kernel under an auxiliary dynamics (related to characteristics). The corresponding result is given in the form of a epigraphical set; the lower envelope of this set is defined as the solution. Once established, the results of this procedure can be fully characterized as follows (details outside of the scope of this book, see, for example, [36]).

Theorem 36 (Generalized Lax-Hopf Formula) *The viability episolution m_c for c as in Definition 31 can be expressed as:*

$$m_c(t, x) = \inf_{(u,s) \in \mathrm{Dom}(f^*) \times \mathbb{R}_{\geq 0}} \Big(c(t - s, x + su) + s f^*(u) \Big). \tag{6.12}$$

Proof See [74].   □

The following Proposition 11 enables it to compute solutions for different input datum (boundary datum, initial datum, datum on trajectories) separately, and then stick all of them together to one solution.

Proposition 11 (Inf-morphism Property) *For $i \in I$ where I is a finite set, let $c_i : \mathrm{Dom} \subset (0, T) \times \mathbb{R} \to \mathbb{R}$ be as in Definition 31. Then, defining*

$$c(t, x) := \min_{i \in I} c_i(t, x), \quad (t, x) \in (0, T) \times X$$

a property known as inf-morphism property holds:

$$\forall (t, x) \in [0, T] \times X, \ m_c(t, x) = \min_{i \in I} m_{c_i}(t, x) \tag{6.13}$$

with m_{c_i} being the solution for the value condition c_i as in Theorem 36.

Proof The proof can be found in [74].   □

6.3.1 Piecewise Affine Initial and Boundary Datum

In this section, we will investigate the generalized Lax-Hopf formula for specific initial and boundary datum, namely for piecewise affine datum. The restriction to this type of initial and boundary datum offers to compute the solution efficiently as a convex optimization problem, or—even more—state the solution explicitly.

On the level of conservation laws this means that we approximate initial and boundary datum piecewise constant. As piecewise constant functions are dense in $L^1(\mathbb{R})$, $L^1(X)$ respectively, we can approximate every initial and boundary flux datum as precise as we want.

6.3.2 Piecewise Affine Initial Datum

Defining piecewise affine initial datum, we recall Eq. (6.6) so that we know that $m_x(0, x) = -u_0(x)$ and as $u_0(x) \in [0, u_{\max}] \ \forall x \in X$ a.e., we define as follows

Definition 32 (Affine Initial Conditions) We consider for $T \in \mathbb{R}_{>0}$ the following affine initial condition $m_0 \colon \mathrm{Dom}(m_0) \subset [0, T] \times X \to \mathbb{R}$:

$$m_0(t, x) = \begin{cases} ax + b & \text{if } x \in [\underline{\alpha}, \overline{\alpha}] \wedge t = 0 \\ +\infty & \text{otherwise,} \end{cases}$$

where $\underline{\alpha}, \overline{\alpha} \in X$ with $\underline{\alpha} \leq \overline{\alpha}$ and $a \in [-u_{\max}, 0], b \in \mathbb{R}$. The effective domain of m_0 is

$$\mathrm{Dom}(m_0) = \{0\} \times [\underline{\alpha}, \overline{\alpha}].$$

Lemma 5 (Lax-Hopf Formula Affine Initial Conditions in Definition 32) *The generalized solution to the Hamilton-Jacobi equation with initial datum as in Definition 32 can be expressed for $(t, x) \in (0, T) \times X$ as*

$$m_{m_0}(t, x) = \begin{cases} \displaystyle\inf_{\substack{u \in \mathrm{Dom}(f^*) \\ \cap\left[\frac{\underline{\alpha}-x}{t}, \frac{\overline{\alpha}-x}{t}\right]}} a(x + tu) + b + tf^*(u) & \mathrm{Dom}(f^*) \cap \left[\frac{\underline{\alpha}-x}{t}, \frac{\overline{\alpha}-x}{t}\right] \neq \emptyset \\ \infty & \text{else.} \end{cases}$$

The solution can actually be stated explicitly for triangular flux functions:

Lemma 6 (Explicit Solution Formula for Definition 32 and Triangular Flux) *For triangular flux function as in Definition 30 the solution formula in Lemma 5 can be stated as*

$$m_{m_0}(t, x)$$
$$= \begin{cases} \begin{cases} a\underline{\alpha} + b + u_c(tv - x) & a + u_c \geq 0 \\ a\overline{\alpha} + b + u_c(tv - x) & a + u_c < 0 \end{cases} & x \in [\max\{A, \underline{\alpha} + wt\}, \min\{B, \overline{\alpha} + wt\}] \\ \infty & \text{else} \end{cases}$$

Proof According to Definition 32 and $\mathrm{Dom}(f^*) = [-v, -w]$ we have for $(t, x) \in (0, T) \times X$ so that $[-\frac{A-x}{v^b}, +\infty) \cap \left[t - \overline{\gamma}, t - \underline{\gamma}\right] \neq \emptyset$ (recall that $a \leq 0$)

$$\inf_{\substack{u \in \left[\frac{\underline{\alpha}-x}{t}, \frac{\overline{\alpha}-x}{t}\right] \cap \\ [-v, -w]}} a \cdot (x + tu) + b + tf^*(u)$$

$$= ax + b + tu_c v + t \cdot \inf_{u \in \left[\frac{\underline{\alpha}-x}{t}, \frac{\overline{\alpha}-x}{t}\right]} (a + u_c)u$$

$$= ax + b + tu_c v + t \cdot \begin{cases} (a + u_c)\frac{\underline{\alpha}-x}{t} & a + u_c \geq 0 \\ (a + u_c)\frac{\overline{\alpha}-x}{t} & a + u_c < 0 \end{cases}$$

$$= \begin{cases} a\underline{\alpha} + b + u_c(tv - x) & a + u_c \geq 0 \\ a\overline{\alpha} + b + u_c(tv - x) & a + u_c < 0 \end{cases}$$

so that we obtain indeed

$$m_{m_0}(t, x) = \begin{cases} \begin{cases} a\underline{\alpha} + b + u_c(tv - x) & a + u_c \geq 0 \\ a\overline{\alpha} + b + u_c(tv - x) & a + u_c < 0 \end{cases} & \left[\frac{\underline{\alpha} - x}{t}, \frac{\overline{\alpha} - x}{t}\right] \cap [-v, -w] \neq \emptyset \\ \infty & \text{else,} \end{cases}$$

which can be reformulated to holds exactly the proposed solution formula. □

6.3.3 Piecewise Affine Left Hand Side Boundary Datum

As the boundary datum for the Hamilton-Jacobi equation is time-integrated flow datum, we define as follows:

Definition 33 (Affine Left Hand Side Boundary Datum) We consider the following upstream boundary datum $\gamma : [0, T] \times X$:

$$\gamma(t, x) = \begin{cases} ct + d & \text{if } t \in [\underline{\gamma}, \overline{\gamma}] \wedge x = A \\ +\infty & \text{otherwise,} \end{cases}$$

where $c \in [0, f_{\max}]$, $d \in \mathbb{R}$ and $\underline{\gamma}, \overline{\gamma} \in [0, T]$ with $\overline{\gamma} - \underline{\gamma} \geq 0$. The effective domain of γ is

$$\text{Dom}(\gamma) = [\underline{\gamma}, \overline{\gamma}] \times \{A\}.$$

Lemma 7 (Lax-Hopf Formula for Affine Left Hand Side b.c.) *The generalized solution to the Hamilton-Jacobi equation with left hand side boundary datum as in Definition 34 can be expressed for* $(t, x) \in [0, T] \times X \setminus \{A\}$ *as*

$$m_\gamma(t, x) = \begin{cases} \inf_{s \in \left[-\frac{A-x}{v^b}, +\infty\right)} c(t - s) + d + sf^*\left(\frac{A-x}{s}\right) & \left[-\frac{A-x}{v^b}, +\infty\right) \cap \left[t - \overline{\gamma}, t - \underline{\gamma}\right] \neq \emptyset \\ \cap\left[t - \overline{\gamma}, t - \underline{\gamma}\right] \\ \infty & else. \end{cases}$$

In case of a triangular flux function, we present the explicit solution formula in the following Lemma 8.

Lemma 8 (Explicit Solution Formula for Lemma 7 and Triangular Flux) *For triangular flux function as in Definition 30 the solution formula in Lemma 7 can be stated for* $t \in [\underline{\gamma}, T]$ *as*

$$
m_\gamma(t, x) = \begin{cases} \begin{cases} u_c(a + vt - x) - \overline{\gamma}(u_c v - c) + d & x \le A + v(t - \overline{\gamma}) \\ \frac{c}{v}(A - x) + ct + d & x \ge A + v(t - \overline{\gamma}) \end{cases} & x \le A + v(t - \underline{\gamma}) \\ \infty & \text{else.} \end{cases}
$$

Proof According to Definition 32, we have for $(t, x) \in (0, T) \times X$, $[-\frac{A-x}{v^b}, +\infty) \cap \left[t - \overline{\gamma}, t - \underline{\gamma}\right] \ne \emptyset$ (recall that $c \in [0, f_{\max}]$)

$$
\inf_{\substack{s \in [-\frac{A-x}{v^b}, +\infty) \\ \cap\left[t - \overline{\gamma}, t - \underline{\gamma}\right]}} c(t - s) + d + s f^* \left(\frac{A-x}{s}\right)
$$

recall that $v^b = v$

$$
= \inf_{\substack{s \in [-\frac{A-x}{v}, +\infty) \\ \cap\left[t - \overline{\gamma}, t - \underline{\gamma}\right]}} c(t - s) + d + s u_c\left(\frac{A-x}{s} + v\right)
$$

$$
= ct + d + u_c(A - x) + \min_{\substack{s \in [-\frac{A-x}{v}, +\infty) \\ \cap\left[t - \overline{\gamma}, t - \underline{\gamma}\right]}} s(u_c v - c)
$$

and also $c \le f_{\max} = f(u_c) = u_c v$

$$
= ct + d + u_c(A - x) + \max\left\{-\frac{A-x}{v}, t - \overline{\gamma}\right\}(u_c v - c)
$$

$$
= \begin{cases} u_c(a + vt - x) - \overline{\gamma}(u_c v - c) + d & x \le A + v(t - \overline{\gamma}) \\ \frac{c}{v}(A - x) + ct + d & x \ge A + v(t - \overline{\gamma}), \end{cases}
$$

from which the conclusion follows.   □

Definition 34 (Affine Right Hand Side Boundary Datum) We consider the following right hand side boundary datum $\beta : [0, T] \times X$:

$$
\beta(t, x) = \begin{cases} et + f & \text{if } t \in [\underline{\beta}, \overline{\beta}] \wedge x = B \\ +\infty & \text{otherwise,} \end{cases} \tag{6.14}
$$

where $e \in [0, f_{max}]$, $f \in \mathbb{R}$ are given as well as $\overline{\beta}, \underline{\beta} \in [0, T]$ with $\overline{\beta} - \underline{\beta} \geq 0$. The effective domain of β is therefore

$$\text{Dom}(\beta) = [\underline{\beta}, \overline{\beta}] \times \{B\}.$$

Lemma 9 (Lax-Hopf Formula for Affine Right Hand Side b.c.) *The generalized solution to the Hamilton-Jacobi equation with right hand side boundary datum as in Definition 34 can be expressed for $(t, x) \in (0, T) \times X \setminus \{B\}$ as*

$$m_\beta(t, x) = \begin{cases} \displaystyle\inf_{\substack{s \in [\frac{B-x}{v^\sharp}, +\infty) \\ \cap [t-\overline{\beta}, t-\underline{\beta}]}} e(t-s) + f + sf^*\left(\frac{B-x}{s}\right) & \left[\frac{B-x}{v^\sharp}, +\infty\right) \cap \left[t-\overline{\beta}, t-\underline{\beta}\right] \neq \emptyset \\ \infty & \text{else.} \end{cases}$$

As in Lemma 8, we will give the solution explicitly for triangular flux function and left hand side boundary datum:

Lemma 10 (Explicit Solution Formula for Lemma 9 and Triangular Flux) *For triangular flux function as in Definition 30 the solution formula in Lemma 9 can be stated for $t \in [\underline{\beta}, T]$ as*

$$m_\beta(t, x)$$

$$= \begin{cases} \begin{cases} u_{max}(B - x) + \frac{x-B}{w}e + et + f & x \leq B + w(t - \overline{\beta}) \\ u_c(B + vt - x) + \overline{\beta}(e - u_c v) + f & x \geq B + w(t - \overline{\beta}) \end{cases} & x \geq B + w(t - \underline{\beta}) \\ \infty & \text{else.} \end{cases}$$

Proof According to Definition 32, we have for $(t, x) \in (0, T) \times X$, $[\frac{B-x}{v^\sharp}, +\infty)$ $\cap \left[t - \overline{\beta}, t - \underline{\beta}\right] \neq \emptyset$ (recall that $e \in [0, f_{max}]$)

$$\inf_{\substack{s \in [\frac{x-B}{v^\sharp}, +\infty) \\ \cap [t-\overline{\beta}, t-\underline{\beta}]}} e(t-s) + f + sf^*\left(\frac{B-x}{s}\right)$$

recall that $v^\sharp = -w$

$$= \inf_{\substack{s \in [\frac{x-B}{w}, +\infty) \\ \cap [t-\overline{\beta}, t-\underline{\beta}]}} e(t-s) + f + su_c\left(\frac{B-x}{s} + v\right)$$

$$= et + f + u_c(B - x) + \min_{\substack{s \in [\frac{x-B}{w}, +\infty) \\ \cap \left[t - \overline{\beta}, t - \underline{\beta}\right]}} s(u_c v - e)$$

and also $c \leq f_{\max} = f(u_c) = u_c v$

$$= et + f + u_c(B - x) + \max\left\{\tfrac{x-B}{w}, t - \overline{\beta}\right\}(u_c v - e)$$

$$= \begin{cases} u_{\max}(B - x) + \tfrac{x-B}{w}e + et + f & x \leq B + w(t - \overline{\beta}) \\ u_c(B + vt - x) + \overline{\beta}(e - u_c v) + f & x \geq B + w(t - \overline{\beta}), \end{cases}$$

from which the conclusion follows. $\qquad\qquad\qquad\qquad\qquad\qquad\qquad\square$

Having stated the solution for any piecewise affine initial datum and boundary datum, we can now move to present a solution formula for unification of such piecewise affine initial and boundary datum.

Definition 35 (The Piecewise Affine Data) Let the sets $I, J, K \subset \mathbb{N}_{\geq 1}$ be the index sets for the piecewise affine initial, left, and right hand side data. We assume that $|I|, |J|, |K| < \infty$. Then, we consider for $(i, j, k) \in I \times J \times K$ the initial, left, and right hand side boundary datum

$$m_{0,i}(t, x) = \begin{cases} a_i x + b_i & \text{if } x \in [\underline{\alpha}_i, \overline{\alpha}_i] \wedge t = 0 \\ +\infty & \text{otherwise} \end{cases}$$

$$\gamma_j(t, x) = \begin{cases} c_j t + d_j & \text{if } t \in [\underline{\gamma}_j, \overline{\gamma}_j] \wedge x = A \\ +\infty & \text{otherwise,} \end{cases}$$

$$\beta_k(t, x) = \begin{cases} e_k t + f_k & \text{if } t \in [\underline{\beta}, \overline{\beta}] \wedge x = B \\ +\infty & \text{otherwise,} \end{cases}$$

with corresponding effective domain and $a \in [-u_{\max}, 0]^{|I|}, b \in \mathbb{R}^{|I|}, \underline{\alpha}, \overline{\alpha} \in X^{|I|}$ with $\underline{\alpha} \leq \overline{\alpha}, c \in [0, f_{\max}]^{|J|}, d \in \mathbb{R}^{|J|}, \underline{\gamma}, \overline{\gamma} \in [0, T]^{|J|}$ with $\underline{\gamma} \leq \overline{\gamma}, e \in [0, f_{\max}]^{|K|}, f \in \mathbb{R}^{|K|}, \underline{\beta}, \overline{\beta} \in [0, T]^{|K|}$ with $\underline{\beta} \leq \overline{\beta}$ the corresponding vectors collecting in their entries the different components of the set of piecewise affine linear initial and boundary data.

By the foregoing Theorems we have for every $(i, j, k) \in I \times J \times K$ the existence of a generalized solution to the corresponding Hamilton-Jacobi equation, just incorporating the specific initial, upstream, downstream, internal data. Each solution is denoted by

$$m_{m_{0,i}}, \quad m_{\gamma_j}, \quad m_{\beta_k}. \qquad\qquad\qquad (6.15)$$

This brings us to the following Theorem, which incorporates all these conditions into one single solution.

Theorem 37 (Solution to the Hamilton-Jacobi Equation for Piecewise Affine Data) *There exists a solution of the Hamilton-Jacobi Equation, incorporating the finite sequences of initial, boundary, and internal data, which we call $c : [0, T] \times X \to \mathbb{R}$. Then, the solution can be determined for $(t, x) \in [0, T] \times X$ by the formula*

$$m(t, x) = \min \left\{ \min_{i \in I} m_{m_{0,i}}(t, x), \ \min_{j \in J} m_{\gamma_j}(t, x), \ \min_{k \in K} m_{\beta_k}(t, x) \right\}, \qquad (6.16)$$

where the involved functions are introduced in Definition 6.15.

Proof The proof is a direct consequence of the inf-morphism property as stated in Proposition 11 in Eq. (6.13) and Definition 35. □

Theorem 38 (Convexity of the Solutions When Changing the Parameters of Initial and Boundary Datum in Definition 35) *For any $(t, x) \in [0, T] \times X$ the solution $m(t, x)$ as defined in Eq. (6.16) is convex with regard to the input variables of initial and boundary datum, i.e., a, b, c, d, e, f in the corresponding dimension.*

Proof We only sketch the proof. Due to the specific construction of the solution being the minimum of the minimum of different functions the solution is convex if each of the contributing solutions in Eq. (6.16) is convex. We start with two initial conditions defined on the same domain, so take according to Definition 32 for given $\underline{\alpha}, \overline{\alpha} \in X : \underline{\alpha} \leq \overline{\alpha}$ as initial datum $m_0(t, x) = ax + b$, $\tilde{m}_0(t, x) = \tilde{a}x + \tilde{b}$ satisfying the constraints in Definition 32 and in particular $(a, \tilde{a}) \in [-u_{\max}, 0]^2$. Then, the corresponding solutions are given in Lemma 6 and applying the definition of convexity, we obtain for $\lambda \in (0, 1)$ and $t \in [0, T]$, $x \in [\max\{A, \underline{\alpha} + wt\}, \min\{B, \overline{\alpha} + wt\}]$

$$m_{\lambda \cdot m_0 + (1-\lambda) \cdot \tilde{m}_0}(t, x)$$

$$= \begin{cases} (\lambda a + (1 - \lambda)\tilde{a})\underline{\alpha} + \lambda b + (1 - \lambda)\tilde{b} + u_c(tv - x), & \lambda a + (1 - \lambda)\tilde{a} + u_c \geq 0 \\ (\lambda a + (1 - \lambda)\tilde{a})\overline{\alpha} + \lambda b + (1 - \lambda)\tilde{b} + u_c(tv - x), & \lambda a + (1 - \lambda)\tilde{a} + u_c < 0 \end{cases}$$

$$= u_c(tv - x) + \lambda \begin{cases} a\underline{\alpha} + b, & \lambda a + (1 - \lambda)\tilde{a} + u_c \geq 0 \\ a\overline{\alpha} + b, & \lambda a + (1 - \lambda)\tilde{a} + u_c < 0 \end{cases}$$

$$+ (1 - \lambda) \begin{cases} \tilde{a}\underline{\alpha} + \tilde{b}, & \lambda a + (1 - \lambda)\tilde{a} + u_c \geq 0 \\ \tilde{a}\overline{\alpha} + \tilde{b}, & \lambda a + (1 - \lambda)\tilde{a} + u_c < 0 \end{cases}$$

$$\leq u_c(tv - x) + \lambda \begin{cases} a\underline{\alpha} + b, & \lambda a + u_c \geq 0 \\ a\overline{\alpha} + b, & \lambda a + u_c < 0 \end{cases}$$

$$+ (1 - \lambda) \begin{cases} \tilde{a}\underline{\alpha} + \tilde{b}, & (1 - \lambda)\tilde{a} + u_c \geq 0 \\ \tilde{a}\overline{\alpha} + \tilde{b}, & (1 - \lambda)\tilde{a} + u_c < 0 \end{cases}$$

$$= \lambda m_{m_0}(t, x) + (1 - \lambda)m_{\tilde{m}_0}(t, x),$$

where the inequality in the previous calculations follows from the fact that

- $\left(a\underline{\alpha} \geq a\overline{\alpha}\right) \wedge \left(\tilde{a}\underline{\alpha} \geq \tilde{a}\overline{\alpha}\right)$ as $a, \tilde{a} \leq 0$
- $\{a \in [-u_{\max}, 0] : \lambda a + (1 - \lambda)\tilde{a} \geq -u_c\} \subseteq \{a \in [-u_{\max}, 0] : \lambda a \geq -u_c\}$
- $\{\tilde{a} \in [-u_{\max}, 0] : \lambda a + (1 - \lambda)\tilde{a} \geq -u_c\} \subseteq \{\tilde{a} \in [-u_{\max}, 0] : (1 - \lambda)\tilde{a} \geq -u_c\}$

for every $\lambda \in [0, 1]$. Similar calculations for the boundary datum using the solution formulae in Lemma 8, Lemma 10 give the convexity with regard to the input parameters of the boundary datum.   □

Remark 17 (Greenshields Flux Function) A similar explicit solution formula can be computed for other flux functions like the quadratic Greenshields [159] flux function $f(x) = vx(u_{\max} - x)$, $x \in [0, u_{\max}]$. In this case the Legendre Fenchel dual f^* will be quadratic as well so that one obtains a quadratic scalar optimization problem with constraints.

As previously stated Theorem 37 gives us a solution formula for any initial and boundary datum where one can also decouple the computation of each part of initial datum and boundary datum. However, for physical relevant solutions (i.e., that there exists the spatial derivative of the solution almost everywhere) one needs to prescribe additional constraints, the so-called compatibility constraints assuring that initial datum and boundary datum fit to each other. For instance, at the space point $(0, 0)$ where initial and boundary datum meet, the datum needs to be continuous over this corner. In addition, assume that we have two parts of piecewise affine linear initial datum as parametrized in Definition 35 so that $\overline{\alpha}_1 = \underline{\alpha}_2$. Then, the corresponding datum does not need to satisfy a continuity assumptions, which would be $a_1\overline{\alpha}_1 + b_1 = a_2\overline{\alpha}_1 + b_2$. Same does not necessarily hold true for the remaining initial and boundary conditions and is the reason why one needs additional compatibility conditions.

6.3.4   Compatibility Conditions

To make sure that compatibility is satisfied so that we indeed obtain solutions of the underlying LWR PDE, we state the following

Theorem 39 (Compatibility Condition) *For the sequence of initial and boundary data as in Definition 35 the data is compatible iff*

$$\forall c, \tilde{c} \in \{m_{0,i} : i \in I\} \cup \{\gamma_j : j \in J\} \cup \{\beta_k : k \in K\}$$

it holds

$$m_c(t, x) \geq \tilde{c}(t, x) \quad \forall (t, x) \in [0, T] \times \overline{X}, \tag{6.17}$$

where m_c is the solution operator for an initial or boundary condition c as discussed in Sect. 6.3.2.

Proof See [74].   □

The compatibility condition might not look very applicable as it states compatibility by just computing the solution and checking then. However, in the case where we can compute the solution explicitly as we have shown in the previous Sect. 6.3.2 for piece-wise affine linear datum, these inequalities can be checked directly. Even more, as the datum is piecewise affine linear Eq. (6.17) does not need to be checked for all $(t, x) \in T \times X$ but only at the boundaries of the domain of each datum and at possible intersections of the corresponding solutions.

6.4   Optimization with Hamilton-Jacobi Equations

In this section, we show for an easy example how the introduced theory and framework in Sect. 6.3 can be used to formulate optimal control problems in a very efficient way as convex (or even linear) optimization problems.

Problem 2 (Problem Considered)   Assume we have at a specific time $T \in \mathbb{R}_{>0}$ measured the road density as $u_T \in L^\infty(X; [0, u_{\max}])$ and the downstream flow $f_B \in L^\infty((0, T); [0, f_{\max}])$. Can we infer the initial state $u_0 \in L^\infty(\mathbb{R})$ and upstream flow $f(u(\cdot, A)) \in L^\infty((0, T))$?

We will address this problem by means of an optimal control problem. As we will take advantage of the previously developed Hamilton-Jacobi theory and generalized solutions (Sect. 6.3), we need to reformulate f_B and u_T as boundary values and end values for the Hamilton-Jacobi equations. Using the relation in Eqs. (6.6) and (6.5) we obtain for the Hamilton-Jacobi equation the following:

Remark 18 (Reformulation in Terms of Hamilton-Jacobi Equations)   The downstream datum and end datum in Problem 2 read for the Hamilton-Jacobi equations for $(t, x) \in [0, T] \times (A, B)$ as

$$h_B(t) := -\int_t^T f_B(s) \, ds, \qquad h_T(x) := \int_x^B u_T(x) \, dx. \tag{6.18}$$

Note that the choice of integral bounds makes the end term compatible with the boundary term in the way that $h_T(B) = 0 = h_B(T)$. This is necessary as the datum we would like to track should be Lipschitz continuous. Although, h_T and

h_B are Lipschitz on their corresponding space/ time domain, they will not satisfy compatibility at the space-time point (T, B).

Now, we are able to formulate the corresponding optimal control problem in terms of Hamilton-Jacobi equations as in Definition 36:

Definition 36 The Optimal Control Problem Considered For $(v, \sigma) \in \mathbb{R}_{>0}^2$ we consider the following constrained minimization problem

$$\inf_{\substack{m_0 \in W^{1,\infty}((X)) \\ h_B W^{1,\infty}((0,T))}} v \|h_B - m(\cdot, B)\|_{L^2((0,T))}^2 + \sigma \|m(T, \cdot) - h_T\|_{L^2(X)}^2,$$

where m is the solution of the Hamilton-Jacobi equation m for initial datum m_0 and left hand side (upstream) boundary datum h_A as stated in Theorem 36 and Proposition 11 for the corresponding value functions.

We will not go into details whether a minimum exists in the chosen functional setup but directly approach the problem in a simplified version: Restricting the flux to a triangular flux as in Definition 30 and the initial and boundary datum to a piecewise affine datum, we can simplify the problem as follows:

Definition 37 (A Finite-Dimensional Optimization Problem for Triangular Flux Function and Piecewise Affine Linear Initial and Left Hand Side Boundary Datum) Chose for the left hand side boundary datum and initial datum as in Eq. (6.15) the finite set I and J with corresponding $\overline{\alpha}, \underline{\alpha} \in X^{|I|}$ so that $\cup_{i \in I}[\underline{\alpha}_i, \overline{\alpha}_i] = \overline{X}$ and $\overline{\gamma}, \underline{\gamma} \in [0, T]^{|J|}$ so that $\cup_{j \in J}[\underline{\gamma}_j, \overline{\gamma}_j] = [0, T]$ a finite-dimensional approximation to Definition 36 reads as

$$\inf_{\substack{a \in [-u_{\max}, 0]^{|I|} \\ b \in \mathbb{R}^{|I|} \\ c \in [0, f_{\max}]^{|J|} \\ d \in \mathbb{R}^{|J|}}} v \|h_B - m(\cdot, B)\|_{L^2((0,T))}^2 + \sigma \|m(T, \cdot) - h_T\|_{L^2(X)}^2$$

$$m(t, x) = \min \left\{ \min_{i \in I} m_{m_{0,i}}(t, x), \ \min_{j \in J} m_{\gamma_j}(t, x) \right\}$$

$m_{m_{0,i}}$ as in Lemma 6 $i \in I$

m_{γ_j} as in Lemma 8 $j \in J$.

As this minimization problem will have results which are not of interest as the corresponding initial and boundary values are not attained we can add a penalization for b, d and compatibility constraints. We then obtain

Theorem 40 (A Convex Optimization Problem) *Adding to the optimization problem in Definition 37 the corresponding compatibility constraints Theorem 39, we still obtain a convex optimization problem in the optimization variables*

$$a, b, c, d.$$

Proof This is a direct consequence of the structure of the compatibility constraints in Theorem 39, and the convexity of the solution in Theorem 38. □

Of course, the previously outlined optimization problem could easily be generalized to more complex situations and the underlying convexity structure could still be applied. In addition, as an explicit solution formula is available, the computation of the minimization problems is fast. Under specific circumstances and more manipulations one can even obtain linear or quadratic and/or mixed-integer programs [61, 74, 221].

6.5 Bibliographical Notes

For general theory and viscosity solutions of Hamilton-Jacobi equation we refer the reader to [94, 95] where the authors consider time-dependent and independent Hamilton-Jacobi equations with a Hamiltonian which can also be explicitly space and time-dependent with Dirichlet boundary conditions and as Cauchy problem. They introduced a notation of solutions, i.e., viscosity solutions for which existence and uniqueness of solutions and stability properties can be obtained (compare also [28]). For a rather comprehensive presentation on optimal control, the related Hamilton-Jacobi-Bellman equations (as optimality conditions) and the named viscosity solutions we refer the reader to [27, 71]. In the book [225] general solutions of Hamilton-Jacobi equations are discussed, in [30] and [134] semicontinuous viscosity solutions. For viability theory we refer to [22, 36] and for Hamilton-Jacobi equations with inequality constraints to [21].

Applications of Hamilton-Jacobi equations and related theory for transportation can be found in [72–74] which are one of the main sources for the latter chapter.

Appendix A
Conservation and Balance Laws and Boundary Value Problems

This appendix is intended to give some background for the reader about the mathematical theory of conservation and balance laws, possibly coupled with boundary data.

A.1 Basic Definitions

A system of balance laws in one space dimension can be written in the form

$$\partial_t u + \partial_x f(u) = g(u), \tag{A.1}$$

where t is the time, x is the space, $u : [0, +\infty[\times \mathbb{R} \to \Omega \subseteq \mathbb{R}^n$ is the *conserved quantity*, $f : \Omega \to \mathbb{R}^n$ is the flux function, and $g : \Omega \to \mathbb{R}^n$ is the source term. In the case $g \equiv 0$, the system (A.1) is called a *system of conservation laws*. Such terminology is justified by the following observation. Integrating (A.1) on an arbitrary space interval $[a, b]$ and assuming for simplicity u is a smooth function, then,

$$\frac{d}{dt} \int_a^b u(t, x)\, dx = \int_a^b \partial_t u(t, x)\, dx = -\int_a^b \partial_x f(u(t, x))\, dx + \int_a^b g(u(t, x))\, dx$$

$$= f(u(t, a)) - f(u(t, b)) + \int_a^b g(u(t, x))\, dx.$$

Thus the variation of the amount of u in the interval $[a, b]$ is related to the quantity of u entering and exiting, respectively, at $x = a$ and $x = b$ and to the quantity of u generated by the source function g.

The flux function f is always supposed to be smooth. Therefore (A.1) can be written, if u is smooth, in the quasi-linear form

A. Bayen et al., *Control Problems for Conservation Laws with Traffic Applications*, PNLDE Subseries in Control 99, https://doi.org/10.1007/978-3-030-93015-8

$$\partial_t u + A(u)\partial_x u = g(u), \tag{A.2}$$

where $A(u) = Df(u)$ is the Jacobian matrix of f at u.

Definition 38 We say that (A.2), or (A.1), is *a hyperbolic system* (respectively, *a strictly hyperbolic system*) if, for every $u \in \Omega$, all the eigenvalues of the Jacobian matrix $A(u)$ are real (respectively, real and distinct).

Example 6 The **Aw–Rascle–Zhang** model for traffic flow is

$$\begin{cases} \partial_t \rho + \partial_x (y - \rho\, p(\rho)) = 0 \\ \partial_t y + \partial_x \left(\frac{y}{\rho}(y - \rho\, p(\rho))\right) = 0, \end{cases}$$

where $\rho = \rho(t, x)$ denotes the average density of cars at time t and at position x, $y = \rho(v + p(\rho))$ is a generalized momentum, v is the velocity of cars, p is a "pressure" term, and $\gamma > 0$ is a fixed parameter. In this situation, $n = 2$ (i.e., the system is composed of $n = 2$ equations), the conserved vector variable u is (ρ, y), and the flux $f(u)$ is given by $\left(y - \rho\, p(\rho), \frac{y}{\rho}(y - \rho\, p(\rho))\right)$. Therefore, the Jacobian matrix A of the flux f is

$$A(u) = \begin{pmatrix} -p(\rho) - \rho\, p(\rho) & 1 \\ -\frac{y^2}{\rho^2} - y\, p'(\rho) & 2\frac{y}{\rho} - p(\rho) \end{pmatrix}$$

and so its eigenvalues are

$$\lambda_1 = \frac{y}{\rho} - p(\rho) - \rho\, p'(\rho), \qquad \lambda_2 = \frac{y}{\rho} - p(\rho).$$

Clearly, if $\rho > 0$ and $p'(\rho) > 0$, then $\lambda_1 < \lambda_2$, and so the system is strictly hyperbolic.

Example 7 The p-**system** model in Eulerian coordinates for gas flow in a tube is

$$\begin{cases} \partial_t \rho + \partial_x q = 0 \\ \partial_t q + \partial_x \left(\frac{q^2}{\rho} + p(\rho)\right) = 0, \end{cases}$$

where $\rho = \rho(t, x)$ denotes the average density of gas at time t and at position x, $q = q(t, x)$ is the linear momentum, and $p = p(\rho)$ is the pressure term. Again, $n = 2$ (i.e., the system is composed of $n = 2$ equations), the conserved vector variable u is (ρ, q), and the flux $f(u)$ is given by $\left(q, \frac{q^2}{\rho} + p(\rho)\right)$. Therefore, the Jacobian matrix A of the flux f is

$$A(u) = \begin{pmatrix} 0 & 1 \\ -\frac{q^2}{\rho^2} + p'(\rho) & 2\frac{q}{\rho} \end{pmatrix}$$

and so its eigenvalues are

$$\lambda_1 = \frac{q}{\rho} - \sqrt{p'(\rho)} \qquad \lambda_2 = \frac{q}{\rho} + \sqrt{p'(\rho)}.$$

Clearly, if $\rho > 0$ and $p'(\rho) > 0$, then $\lambda_1 < \lambda_2$, and so the system is strictly hyperbolic.

Example 8 The **Saint-Venant** or **shallow water** equations for the description of open channels are

$$\begin{cases} \partial_t H + \partial_x (HV) = 0 \\ \partial_t V + \partial_x \left(\frac{V^2}{2} + gH \right) = 0, \end{cases}$$

where $H = H(t, x)$ denotes the water level at time t and at position x, $V = V(t, x)$ is the water velocity, and g is the gravitation constant. In this situation, $n = 2$ (i.e., the system is composed of $n = 2$ equations), the conserved vector variable u is (H, V), and the flux $f(u)$ is given by $\left(HV, \frac{V^2}{2} + gH \right)$. Therefore, the Jacobian matrix A of the flux f is

$$A(u) = \begin{pmatrix} V & H \\ g & V \end{pmatrix}$$

and so its eigenvalues are

$$\lambda_1 = V - \sqrt{gH} \qquad \lambda_2 = V + \sqrt{gH}.$$

Clearly, if $H > 0$, then $\lambda_1 < \lambda_2$, and so the system is strictly hyperbolic.

A.2 BV Functions

Consider an interval $J \subset \mathbb{R}$ and a map $u : J \mapsto \mathbb{R}$. The *total variation* of u is defined as

$$\mathrm{TV}(u) \doteq \sup \left\{ \sum_{i=1}^{N} |u(x_j) - u(x_{j-1})| \right\}, \tag{A.3}$$

where the supremum is taken over all $N \geq 1$ and all $(N + 1)$-tuples of points $x_j \in J$ such that $x_0 < x_1 < \ldots < x_N$. If the total variation of u is finite, then we write $u \in \mathbf{BV}(J; \mathbb{R})$. Specific properties of $\mathbf{BV}$ functions used in this chapter are presented below; for a proof, see [51, Section 2.4].

Lemma 11 *Let* $u :]a, b[\mapsto \mathbb{R}^n$ *have bounded variation. Then, for every* $x \in]a, b[$, *the left and right limits*

$$u(x-) \doteq \lim_{y \mapsto x-} u(y), \quad u(x+) \doteq \lim_{y \mapsto x+} u(y)$$

are well defined. Moreover, u has at most countably many points of discontinuity.

The following lemma concerns piecewise constant approximability of $\mathbf{BV}$ functions.

Lemma 12 *Let* $u : \mathbb{R} \mapsto \mathbb{R}^n$ *be right continuous with bounded variation. Then, for every* $\epsilon > 0$, *there exists a piecewise constant function* v *such that*

$$\mathrm{TV}\,(v) \leq \mathrm{TV}\,(u)\,, \quad \|v - u\|_{\mathbf{L}^\infty} \leq \epsilon.$$

If, in addition,

$$\int_{-\infty}^{0} |u(x) - u(-\infty)| \ \mathrm{d}x + \int_{0}^{+\infty} |u(x) - u(+\infty)| \ \mathrm{d}x < +\infty,$$

then one can find v with the additional property

$$\|u - v\|_{\mathbf{L}^1} \leq \epsilon.$$

The space of $\mathbf{BV}$ functions and its closure in $\mathbf{L}^1$ are at the center of well-posedness results for conservation laws using wave-front tracking methods and other approximation schemes.

A.3   The Method of Characteristics

In this section, we briefly describe the method of characteristics for the Cauchy problem

$$\begin{cases} \partial_t u + a(u)\partial_x u = g(u), \\ u(0, x) = u_0(x), \end{cases} \tag{A.4}$$

where $a = a(u)$ and $g = g(u)$ are given smooth functions, and $\bar{u}$ is a given initial condition. The idea of this method consists in finding one-dimensional curves, along which the solution to (A.4) can be explicitly computed. In this way, a first order partial differential equation can be solved using ordinary differential equations.

For every $y \in \mathbb{R}$, define the functions $t \mapsto X_y(t)$, $t \mapsto U_y(t)$ as the solutions to the system

$$
\begin{cases}
\frac{d}{dt} X_y(t) = a(U_y(t)), \\
\frac{d}{dt} U_y(t) = g(U_y(t)), \\
X_y(0) = y, \\
U_y(0) = u_0(y).
\end{cases}
\tag{A.5}
$$

The curves $t \mapsto X_y(t)$ are called characteristics, while $U_y(t)$ represent the solution u to (A.4) along the characteristics starting at the point $(0, y)$. If the map

$$
(t, y) \mapsto (t, X_y(t))
\tag{A.6}
$$

is invertible, then the inverse function $(t, X) \mapsto (t, y_X(t))$ allows to express the solution u to the Cauchy problem (A.4) in the form

$$
u(t, x) = U_{y_x(t)}(t).
\tag{A.7}
$$

Example 9 Consider the inviscid scalar Burgers' equation

$$
\partial_t u + u \, \partial_x u = 0
\tag{A.8}
$$

with the initial condition

$$
u_0(x) = \begin{cases} 1 - |x|, & \text{if } x \in [-1, 1], \\ 0, & \text{otherwise,} \end{cases}
\tag{A.9}
$$

see Fig. A.1. System (A.5) becomes

$$
\begin{cases}
\frac{d}{dt} X_y(t) = U_y(t), \\
\frac{d}{dt} U_y(t) = 0, \\
X_y(0) = y, \\
U_y(0) = u_0(y),
\end{cases}
\tag{A.10}
$$

and its solution is

$$
\begin{cases}
X_y(t) = y + u_0(y) \, t \\
U_y(t) = u_0(y).
\end{cases}
\tag{A.11}
$$

Therefore,

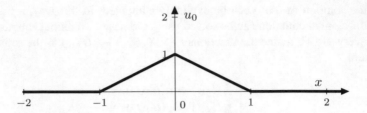

Fig. A.1 The initial datum u_0 of Example 9

Fig. A.2 The characteristic
curves for the Cauchy
problem in Example 9

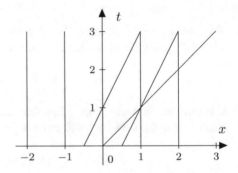

$$X_y(t) = \begin{cases} y, & \text{if } y < -1, \\ y + (1 + y)t, & \text{if } -1 < y < 0, \\ y + (1 - y)t, & \text{if } 0 < y < 1, \\ y, & \text{if } y > 1, \end{cases} \qquad (A.12)$$

see Fig. A.2. We notice that, if $t < 1$, then the characteristic lines do not intersect
and the function

$$u(t, x) = \begin{cases} \frac{x+1}{t+1}, & \text{if } -1 \le x < t, \\ \frac{1-x}{1-t}, & \text{if } t < x \le 1, \\ 0, & \text{otherwise} \end{cases}$$

provides a classical solution to the Cauchy problem (A.8)–(A.9). If $t > 1$, then
different characteristic lines intersect and therefore a classical solution does not
exist. In Fig. A.3, the profiles of the solution u at times $t = 0.5$, $t = 0.75$, and
$t = 0.95$ are plotted.

Example 10 Let us consider the inviscid scalar Burgers' equation

$$\partial_t u + u \, \partial_x u = 0,$$

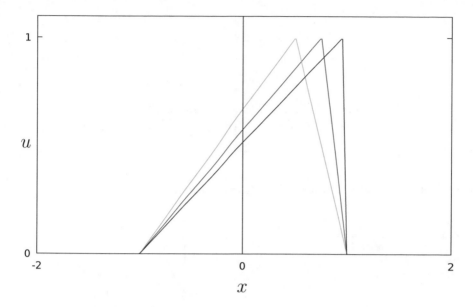

Fig. A.3 Solution to Burgers' equation of Example 9 at times $t = 0.5$, $t = 0.75$, and $t = 0.95$

with the initial condition $u(0, x) = u_0(x) = \frac{1}{1+x^2}$, see Fig. A.4. System (A.4) has the solution

$$\begin{cases} X_y(t) = y + \frac{1}{1+y^2} t, \\ U_y(t) = \frac{1}{1+y^2}. \end{cases} \tag{A.13}$$

Since $U_y(t)$ does not depend on t, then $u(t, x)$ is constant along the lines in the (t, x)-plane described in the parametric form as

$$t \mapsto \left(t, y + \frac{t}{1 + y^2} \right);$$

see Fig. A.5. Moreover, along such curves, the value of u is $\frac{1}{1+y^2}$.

In Fig. A.5, it is clear that the characteristic curves intersect together. More precisely, there exists a time $T = \frac{8}{\sqrt{27}}$ with the following property. For $t < T$, the characteristic lines do not intersect together and so a classical solution exists, but for $t > T$ different characteristics intersect, showing that a classical solution cannot exist for $t \geq T$; see Fig. A.5. In Fig. A.6, the profiles of the solution u to (A.4) at times $t = 0.5$, $t = 1$, and $t = 1.5$ are plotted.

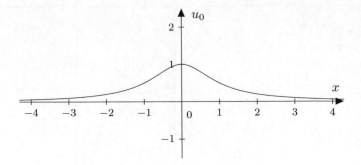

Fig. A.4 The initial datum u_0 of the Cauchy problem in Example 10

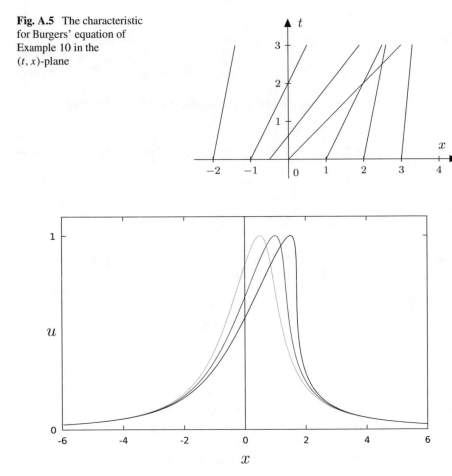

Fig. A.5 The characteristic
for Burgers' equation of
Example 10 in the
(t, x)-plane

Fig. A.6 Solution to Burgers' equation of Example 10 at times $t = 0.5$, $t = 1$, and $t = 1.5$

A.4 Weak Solutions

As noted in Examples 9 and 10, classical solutions to hyperbolic conservation laws do not exist in general for all times $t \geq 0$. This happens independently of the regularity of the initial condition. More precisely, when two different characteristic curves collide, the solution becomes discontinuous. Hence we must deal with the concept of weak or distributional solutions.

Definition 39 Fix $T \in [0, +\infty]$. A function $u \in \mathbf{L}_{loc}^1([0, T] \times \mathbb{R}; \mathbb{R}^n)$ is a weak solution to the balance law

$$\partial_t u + \partial_x f(u) = g(u)$$

if, for every function $\psi \in \mathbf{C}_c^1(]0, T[\times \mathbb{R}; \mathbb{R}^n)$, it holds

$$\int_0^T \int_{\mathbb{R}} (u \cdot \partial_t \psi + f(u) \cdot \partial_x \psi + g(u) \cdot \psi) \, dx \, dt = 0. \tag{A.14}$$

The usual definition of weak solution for a Cauchy problem is the following.

Definition 40 Fix $u_0 \in \mathbf{L}_{loc}^1(\mathbb{R}; \mathbb{R}^n)$ and $T \in [0, +\infty]$. A function $u : [0, T] \times \mathbb{R} \to \mathbb{R}^n$ is a weak solution to the Cauchy problem

$$\begin{cases} \partial_t u + \partial_x f(u) = g(u), \\ u(0, x) = u_0(x), \end{cases} \tag{A.15}$$

if u is continuous as a function from $[0, T]$ into $\mathbf{L}_{loc}^1$, if u is a weak solution to $\partial_t u + \partial_x f(u) = g(u)$, according to Definition 39, and $\lim_{t \to 0^+} u(t, x) = \bar{u}(x)$ in $\mathbf{L}^1(\mathbb{R})$, which means

$$\lim_{t \to 0^+} \|u(t, \cdot) - u_0\|_{\mathbf{L}^1} = 0. \tag{A.16}$$

Weak solutions to conservation laws can have discontinuities, such as shocks or contact discontinuities. The next result gives a necessary and sufficient condition, usually called the Rankine–Hugoniot condition, for piecewise constant functions to be weak solutions. It gives a condition on discontinuities for weak solutions to (A.1) relating the right and left states with the "speed" λ of the discontinuity.

Proposition 12 *Fix time $T > 0$ and constant states $u_1, u_2 \in \mathbb{R}^n$. The function*

$$u(t, x) = \begin{cases} u_1, & \text{if } x < \lambda t, \\ u_2, & \text{if } x > \lambda t \end{cases} \tag{A.17}$$

is a weak solution to $\partial_t u + \partial_x f(u) = 0$ if and only if $\lambda \in \mathbb{R}$ satisfies the Rankine–Hugoniot condition

$$f(u_1) - f(u_2) = \lambda\,(u_1 - u_2)\,. \tag{A.18}$$

Note that (A.18) is indeed a system of n scalar equations. In the case $n = 1$, it can be written in the form

$$\lambda = \frac{f(u_1) - f(u_2)}{u_1 - u_2},$$

provided $u_1 \neq u_2$.

Proof Here we only prove that if u is a weak solution, then the Rankine–Hugoniot condition holds. The converse implication can be deduced in a similar way, and hence we omit it.

Assume therefore that u is a weak solution to $\partial_t u + \partial_x f(u) = 0$. Fix $\psi : \mathbb{R}^2 \to \mathbb{R}^n$ a $\mathbf{C}^1$ function with compact support, contained in a compact and connected set K with smooth boundary, and consider the vector field $\Phi : \mathbb{R}^2 \to \mathbb{R}^2$, defined by

$$\Phi(t, x) = (u(t, x) \cdot \psi(t, x),\ f(u(t, x)) \cdot \psi(t, x))\,.$$

Moreover, define the sets

$$K^+ = K \cap \{x > \lambda t\}, \qquad K^- = K \cap \{x < \lambda t\}, \qquad K^0 = K \cap \{x = \lambda t\};$$

see Fig. A.7. Denote by $\mathbf{n}_+$ and $\mathbf{n}_-$, respectively, the outward normal to K^+ and K^- at points of the boundaries ∂K^+ and ∂K^- of K^+ and K^-. Note that, for points satisfying $x = \lambda t$, the expressions for $\mathbf{n}_+$ and $\mathbf{n}_-$ are given by

$$\mathbf{n}_+ = -\mathbf{n}_- = \frac{1}{\sqrt{\lambda^2 + 1}}(\lambda, -1).$$

The Divergence Theorem, applied to Φ on K^+, implies that

$$\int_{K^+} \mathrm{div}\,\Phi(t, x)\,\mathrm{d}x\,\mathrm{d}t = \int_{\partial K^+} \mathbf{n}_+(t, x) \cdot \Phi(t, x)\,\mathrm{d}\sigma = \int_{K^0} \mathbf{n}_+(t, x) \cdot \Phi(t, x)\,\mathrm{d}\sigma$$

$$= \frac{1}{\sqrt{\lambda^2 + 1}} \int_a^b (\lambda u(t, \lambda t+) - f(u(t, \lambda t+))) \cdot \psi(t, \lambda t+)\,\mathrm{d}t,$$

where

$$\begin{aligned} a &= \inf\{t \geq 0 : (t, x) \in K^0,\ \exists x \in \mathbb{R}\} \\ b &= \sup\{t \geq 0 : (t, x) \in K^0,\ \exists x \in \mathbb{R}\}\,. \end{aligned} \tag{A.19}$$

Moreover, the Divergence Theorem, applied to Φ on K^-, implies that

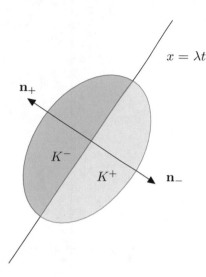

Fig. A.7 Region where the divergence theorem is applied

$$\int_{K^-} \operatorname{div} \Phi(t, x) \, dx \, dt = -\frac{1}{\sqrt{\lambda^2 + 1}} \int_a^b (\lambda u(t, \lambda t+) - f(u(t, \lambda t+))) \cdot \psi(t, \lambda t+) \, dt.$$

The fact that u is weak solution to $\partial_t u + \partial_x f(u) = 0$ implies that

$$0 = \int_K \operatorname{div} \Phi(t, x) \, dx \, dt = \int_{K^+} \operatorname{div} \Phi(t, x) \, dx \, dt + \int_{K^-} \operatorname{div} \Phi(t, x) \, dx \, dt,$$

and so

$$0 = \int \left[\lambda \left(u^+ - u^- \right) - \left(f(u^+) - f(u^-) \right) \right] \cdot \psi(t, \lambda t) \, dt.$$

The arbitrariness of the function ψ implies that

$$\lambda \left(u^+ - u^- \right) = f(u^+) - f(u^-),$$

i.e., λ satisfies condition (A.18).

$\square$

The notion of weak solution for conservation laws does not guarantee uniqueness. The next example shows that there are infinitely many weak solutions for the same Cauchy problem. Therefore, to obtain a unique solution for a Cauchy problem, the notion of weak solution must be supplemented with admissibility conditions, possibly motivated by physical considerations.

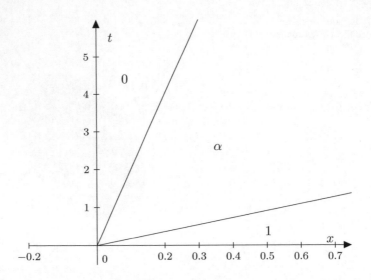

Fig. A.8 The solution u_α of Example 11 in the (t, x)-plane

Example 11 Let us consider the scalar Burgers equation

$$\partial_t u + \partial_x \left(\frac{u^2}{2} \right) = 0,$$

with the initial condition

$$u(0, x) = u_0(x) = \begin{cases} 1, & \text{if } x \geq 0, \\ 0, & \text{if } x < 0. \end{cases}$$

For every $0 < \alpha < 1$, the function $u_\alpha : [0, +\infty[\times \mathbb{R} \to \mathbb{R}$ defined by

$$u_\alpha(t, x) = \begin{cases} 0, & \text{if } x < \frac{\alpha t}{2}, \\ \alpha, & \text{if } \frac{\alpha t}{2} \leq x < \frac{(1+\alpha)t}{2}, \\ 1, & \text{if } x \geq \frac{(1+\alpha)t}{2} \end{cases}$$

is a weak solution to the Burgers equation; see Fig. A.8. Indeed, u_α has two discontinuities along the lines $x = \alpha t/2$ and $x = (1 + \alpha)t/2$; i.e., there are two shocks travelling with speeds $\lambda_1 = \alpha/2$ and $\lambda_2 = (1+\alpha)/2$. Since the flux function is $f(u) = \frac{u^2}{2}$, the corresponding Rankine–Hugoniot conditions

$$f(\alpha) - f(0) = \frac{\alpha^2}{2} - 0 = \frac{\alpha}{2} (\alpha - 0) = \lambda_1(\alpha - 0)$$

and

$$f(1) - f(\alpha) = \frac{1}{2} - \frac{\alpha^2}{2} = \frac{1+\alpha}{2}(1-\alpha) = \lambda_2(1-\alpha)$$

hold.

A.5 Entropy Admissible Solutions

As shown in Example 11, there are situations of non-uniqueness for weak solutions to Riemann and Cauchy problems. Hence additional conditions are necessary in order to isolate a unique solution. A possible way is to choose the solution, which minimizes the entropy dissipation. The next definition introduces therefore the concept of entropy, entropy flux, and entropy production.

Definition 41 A convex function $\eta \in \mathbf{C}^1(\mathbb{R}^n; \mathbb{R})$ is an entropy for (A.1) if there exists a function $q \in \mathbf{C}^1(\mathbb{R}^n; \mathbb{R})$ such that

$$\nabla q(u) = \nabla \eta(u) \cdot \nabla f(u) \qquad (A.20)$$

for every $u \in \mathbb{R}^n$. The function q is called an entropy flux for η and the pair (η, q) is called an entropy–entropy flux pair for (A.1). Finally, the function $h(u) = \nabla \eta(u) \cdot g(u)$ is said entropy production.

In the scalar case, i.e., when $n = 1$, Eq. (A.20) can be written in the form

$$q'(u) = \eta'(u) f'(u). \qquad (A.21)$$

Therefore, for every $\mathbf{C}^1$ convex function η, there exist infinitely many entropy fluxes

$$q(u) = \int_{\bar{u}}^{u} \eta'(s) f'(s) \, ds,$$

where $\bar{u} \in \mathbb{R}$ is arbitrary.

The definition of entropy admissible solution is the following one.

Definition 42 A weak solution $u = u(t, x)$ to (A.1) is said entropy admissible if, for every entropy–entropy flux pair (η, q) and for every $\psi \in \mathbf{C}_c^1(]0, T[\times\mathbb{R}; \mathbb{R})$ satisfying $\psi(t, x) \geq 0$ for all $(t, x) \in]0, T[\times\mathbb{R}$, it holds

$$\int_0^T \int_{\mathbb{R}} (\eta(u)\partial_t \psi + q(u)\partial_x \psi + h(u)\psi) \, dx \, dt \geq 0, \qquad (A.22)$$

where $h(u) = \nabla \eta(u) \cdot g(u)$ is the entropy production.

Example 12 Consider the function

$$u(t, x) = \begin{cases} 0, & \text{if } x < \frac{t}{2}, \\ 1, & \text{if } x \geq \frac{t}{2}, \end{cases}$$

which is a weak solution to the scalar Burgers equation $\partial_t u + \partial_x \left(\frac{u^2}{2} \right) = 0$, see also Example 11.

We show here that u is not entropy admissible, according to Definition 42. To this aim, let us consider the convex entropy $\eta(u) = u^2$ and the corresponding entropy flux $q(u) = \frac{2u^3}{3}$. Fix a test function $\psi \in C_c^1 (]0, T[\times\mathbb{R}; [0, +\infty[)$ such that $\psi \left(\frac{T}{2}, \frac{T}{4} \right) > 0$ and define the sets

$$\Omega_l = \left\{ (t, x) \in [0, T] \times \mathbb{R} : x < \frac{t}{2} \right\} \qquad \Omega_r = \left\{ (t, x) \in [0, T] \times \mathbb{R} : x > \frac{t}{2} \right\}.$$

We have that

$$\int_0^T \int_{\mathbb{R}} (\eta(u(t, x))\partial_t \psi(t, x) + q(u(t, x))\partial_x \psi(t, x)) \ dx \, dt$$

$$= \int \int_{\Omega_l \cup \Omega_r} (\eta(u(t, x))\partial_t \psi(t, x) + q(u(t, x))\partial_x \psi(t, x)) \ dx \, dt$$

$$= \int \int_{\Omega_l} (\eta(0)\partial_t \psi(t, x) + q(0)\partial_x \psi(t, x)) \ dx \, dt$$

$$+ \int \int_{\Omega_r} (\eta(1)\partial_t \psi(t, x) + q(1)\partial_x \psi(t, x)) \ dx \, dt$$

$$= \int \int_{\Omega_r} \text{div} \left(\psi(t, x), \frac{2}{3}\psi(t, x) \right) \ dx \, dt.$$

Applying the Divergence Theorem, we deduce that

$$\int \int_{\Omega_r} \text{div} \left(\psi(t, x), \frac{2}{3}\psi(t, x) \right) \ dx \, dt = \int_0^T \left(1, \frac{2}{3} \right) \cdot \left(\frac{1}{\sqrt{5}}, \frac{-2}{\sqrt{5}} \right) \sqrt{5}\psi \left(t, \frac{t}{2} \right) dt$$

$$= -\frac{1}{3} \int_0^T \psi \left(t, \frac{t}{2} \right) dt < 0$$

and so

$$\int_0^T \int_{\mathbb{R}} (\eta(u(t, x))\partial_t \psi(t, x) + q(u(t, x))\partial_x \psi(t, x)) \ dx \, dt < 0.$$

Consequently, (A.22) does not hold, proving that u is not entropy admissible.

Example 13 Consider the function

$$u(t, x) = \begin{cases} 1, & \text{if } x < \frac{t}{2}, \\ 0, & \text{if } x \geq \frac{t}{2}, \end{cases}$$

which is a weak solution to the scalar Burgers' equation $\partial_t u + \partial_x \left(\frac{u^2}{2} \right) = 0$, see also Example 11.

We show here that u is entropy admissible, according to Definition 42. To this aim, consider an arbitrary convex entropy $\eta(u)$ and the corresponding entropy flux $q(u)$, fix a test function $\psi \in C_c^1 (]0, T[\times \mathbb{R}; [0, +\infty))$, and define the sets

$$\Omega_l = \left\{ (t, x) \in [0, T] \times \mathbb{R} : x < \frac{t}{2} \right\} \qquad \Omega_r = \left\{ (t, x) \in [0, T] \times \mathbb{R} : x > \frac{t}{2} \right\}.$$

We have that

$$\int_0^T \int_{\mathbb{R}} (\eta(u(t, x)) \partial_t \psi(t, x) + q(u(t, x)) \partial_x \psi(t, x)) \; dx \, dt$$

$$= \int \int_{\Omega_l \cup \Omega_r} (\eta(u(t, x)) \partial_t \psi(t, x) + q(u(t, x)) \partial_x \psi(t, x)) \; dx \, dt$$

$$= \int \int_{\Omega_l} (\eta(1) \partial_t \psi(t, x) + q(1) \partial_x \psi(t, x)) \; dx \, dt$$

$$+ \int \int_{\Omega_r} (\eta(0) \partial_t \psi(t, x) + q(0) \partial_x \psi(t, x)) \; dx \, dt$$

$$= \int \int_{\Omega_l} \text{div} \, (\eta(1) \psi(t, x), q(1) \psi(t, x)) \; dx \, dt$$

$$+ \int \int_{\Omega_r} \text{div} \, (\eta(0) \psi(t, x), q(0) \psi(t, x)) \; dx \, dt.$$

Applying the Divergence Theorem, we deduce that

$$\int \int_{\Omega_l} \text{div} \, (\eta(1) \psi(t, x), q(1) \psi(t, x)) \; dx \, dt$$

$$= \int_0^T (\eta(1), q(1)) \cdot \left(\frac{-1}{\sqrt{5}}, \frac{2}{\sqrt{5}} \right) \frac{\sqrt{5}}{2} \psi \left(t, \frac{t}{2} \right) \, dt$$

$$= \left(q(1) - \frac{\eta(1)}{2} \right) \int_0^T \psi \left(t, \frac{t}{2} \right) \, dt$$

and

$$\int\int_{\Omega_r} \text{div}\,(\eta(0)\psi(t,x), q(0)\psi(t,x))\;dx\,dt$$

$$= \int_0^T (\eta(0), q(0)) \cdot \left(\frac{1}{\sqrt{5}}, \frac{-2}{\sqrt{5}}\right) \frac{\sqrt{5}}{2} \psi\left(t, \frac{t}{2}\right) dt$$

$$= \left(\frac{\eta(0)}{2} - q(0)\right) \int_0^T \psi\left(t, \frac{t}{2}\right) dt.$$

Consequently, we deduce that

$$\int_0^T \int_{\mathbb{R}} (\eta(u(t,x))\partial_t\,\psi(t,x) + q(u(t,x))\partial_x\,\psi(t,x))\;dx\,dt$$

$$= \frac{1}{2} (\eta(0) - \eta(1) + 2q(1) - 2q(0)) \int_0^T \psi\left(t, \frac{t}{2}\right) dt.$$

Using (A.20), we have that $q'(u) = u\eta'(u)$ for every u and so

$$q(1) - q(0) = \int_0^1 u\eta'(u)\,du = \eta(1) - \int_0^1 \eta(u)\,du;$$

hence

$$\frac{1}{2} (\eta(0) - \eta(1) + 2q(1) - 2q(0)) = \frac{\eta(0) + \eta(1)}{2} - \int_0^1 \eta(u)\,du.$$

Since η is convex, then

$$\eta(u) \leq \eta(0) + (\eta(1) - \eta(0))\,u$$

for every $u \in [0, 1]$ and so

$$\int_0^1 \eta(u)\,du \leq \eta(0) + (\eta(1) - \eta(0)) \int_0^1 u\,du = \frac{\eta(0) + \eta(1)}{2}.$$

Finally, the positivity of the function ψ implies that

$$\int_0^T \int_{\mathbb{R}} (\eta(u(t,x))\partial_t\,\psi(t,x) + q(u(t,x))\partial_x\,\psi(t,x))\;dx\,dt \geq 0,$$

proving that u is entropy admissible.

A.5.1 Kružkov Entropy Condition

In the scalar case, i.e., when $n = 1$, it is possible to consider entropies which are continuous but not continuously differentiable. In this setting, one obtains an entropy condition, which is known as the Kružkov entropy condition; see [198].

For each $k \in \mathbb{R}$, consider the convex entropy function $\eta_k(u) = |u - k|$, which is not differentiable at $u = k$, and consider the corresponding entropy flux $q_k(u) = \mathrm{sgn}\,(u - k)\,(f(u) - f(k))$.

Definition 43 A weak solution $u = u(t, x)$ to the scalar equation (A.1) satisfies the Kružkov entropy admissibility condition if

$$\int_0^T \int_{\mathbb{R}} \left[|u - k| \, \partial_t \, \psi + \mathrm{sgn}(u - k)\,(f(u) - f(k))\, \partial_x \, \psi + \mathrm{sgn}(u - k)g(u)\psi \right] \mathrm{d}x \, \mathrm{d}t \geq 0$$

for every $k \in \mathbb{R}$ and for every $\psi \in \mathbf{C}_c^1 \left( ]0, T[\times \mathbb{R}; \mathbb{R}^+ \right)$.

A.6   The Riemann Problem

In this section, we briefly describe the entropy admissible solutions to Riemann problems. Let $\Omega \subset \mathbb{R}^n$ be an open set, let $f : \Omega \to \mathbb{R}^n$ be a smooth flux, and consider the strictly hyperbolic system of conservation laws

$$\partial_t u + \partial_x f(u) = 0. \tag{A.23}$$

Definition 44 A Riemann problem for (A.23) is the Cauchy problem (A.15), where the initial condition u_0 has the form

$$u_0(x) := \begin{cases} u^-, & \text{if } x < 0, \\ u^+, & \text{if } x > 0, \end{cases} \tag{A.24}$$

with $u^-, u^+ \in \Omega$.

In view of Definition 44, we describe the solution to the following Riemann problem:

$$\begin{cases} \partial_t u + \partial_x f(u) = 0 \\ u(0, x) = u_0(x) = \begin{cases} u^- & \text{if } x < 0 \\ u^+ & \text{if } x > 0, \end{cases} \end{cases} \tag{A.25}$$

where $u^-, u^+ \in \Omega$. We consider the scalar case, i.e., $n = 1$, and the system case, i.e., $n > 1$, separately.

A.6.1   The Scalar Case

Here we assume that $n = 1$ and that $\Omega = \mathbb{R}$ for simplicity. The case of Ω a real interval is straightforward. Let $f : \mathbb{R} \to \mathbb{R}$ be a smooth flux function, $u^-, u^+ \in \mathbb{R}$, $u^- \neq u^+$. We treat two symmetric situations: the strictly convex and the strictly concave flux.

A.6.1.1   The Riemann Problem for a Strictly Convex Flux

Assume that the flux function $f : \mathbb{R} \to \mathbb{R}$ is strictly convex. The solution to (A.25) depends on the following two possibilities.

$u^- < u^+$:   In this case, $f'(u^-) \leq f'(u^+)$ and so the characteristic lines, starting at $t = 0$, do not intersect together. Therefore (A.25) produces a rarefaction wave and its solution has the form

$$u(t, x) = \begin{cases} u_L, & \text{if } x < f'(u^-)t, \\ g\left(\frac{x}{t}\right), & \text{if } f'(u^-)t < x < f'(u^+)t, \\ u_R, & \text{if } x > f'(u^+)t, \end{cases} \qquad (A.26)$$

where g denotes the inverse map of the derivative f' of the flux; see Fig. A.9. Note that the assumptions on f do imply that f' is injective and so invertible.

$u^- > u^+$:   In this case, $f'(u^-) \geq f'(u^+)$ and so the characteristic curves intersect. Therefore, (A.25) produces a shock wave of the form

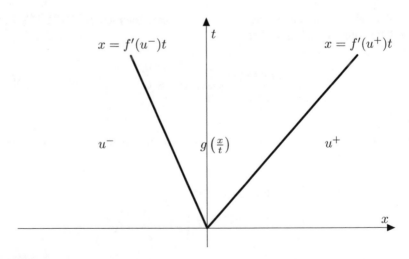

Fig. A.9 The solution (A.26) to the Riemann problem (A.25) with strictly convex flux function in the case $u^- < u^+$ on the (t, x) space

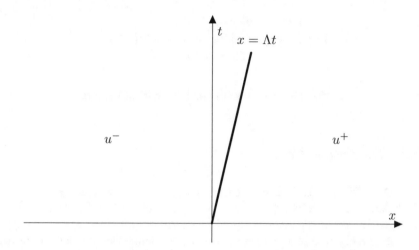

Fig. A.10 The solution (A.27) to the Riemann problem (A.25) with strictly convex flux function in the case $u^- > u^+$ on the (t, x) space

$$u(t, x) = \begin{cases} u^-, & \text{if } x < \lambda t, \\ u^+, & \text{if } x > \lambda t, \end{cases} \tag{A.27}$$

where $\lambda = \frac{f(u^-)-f(u^+)}{u^--u^+}$ is the speed of the jump, given by the Rankine–Hugoniot condition, see Fig. A.10.

A.6.1.2   The Riemann Problem for a Concave Flux

Assume that the flux function $f : \mathbb{R} \to \mathbb{R}$ is strictly concave. The solution to (A.25) depends on the following possibilities:

$u^- < u^+$:   In this case, $f'(u^-) \geq f'(u^+)$ and so there exist two different characteristics, which intersect together. Therefore, (A.25) produces a shock wave and its solution has the form

$$u(t, x) = \begin{cases} u^-, & \text{if } x < \lambda t, \\ u^+, & \text{if } x > \lambda t, \end{cases} \tag{A.28}$$

where $\lambda = \frac{f(u^-)-f(u^+)}{u^--u^+}$ is the speed of the jump, given by the Rankine–Hugoniot condition.

$u^- > u^+$:   In this case, $f'(u^-) \leq f'(u^+)$ and so the characteristic lines, starting from $t = 0$, do not intersect together. Therefore, (A.25) produces a rarefaction wave of the form

$$u(t, x) = \begin{cases} u^-, & \text{if } x < f'(u^-)t, \\ g\left(\frac{x}{t}\right), & \text{if } f'(u^-)t < x < f'(u^+)t, \\ u^+, & \text{if } x > f'(u^+)t, \end{cases} \tag{A.29}$$

where g denotes the inverse map of the derivative f' of the flux.

A.6.2   The System Case

Let $f : \Omega \to \mathbb{R}^n$, $n > 1$, be a smooth flux function, $u^-, u^+ \in \Omega$, $u^- \neq u^+$.

As before, we denote by $A(u)$ the Jacobian matrix of the flux f and with $\lambda_1(u) < \cdots < \lambda_n(u)$ the ordered n eigenvalues of the matrix $A(u)$. Let $\{r_1(u), \ldots, r_n(u)\}$ be a basis of right eigenvectors. For $i \in \{1, \cdots, n\}$, we define the directional derivative of $\lambda_i(u)$ in the direction of $r_i(u)$ as

$$\nabla \lambda_i(u) \cdot r_i(u) := \lim_{\varepsilon \to 0^+} \frac{\lambda_i(u + \varepsilon r_i(u)) - \lambda_i(u)}{\varepsilon}.$$

We need to introduce the definition of genuinely nonlinear characteristic field and of linearly degenerate one.

Definition 45 The i-th characteristic field ($i \in \{1, \cdots, n\}$) is said genuinely nonlinear if

$$\nabla \lambda_i(u) \cdot r_i(u) \neq 0 \qquad \forall u \in \Omega.$$

The i-th characteristic field ($i \in \{1, \cdots, n\}$) is said linearly degenerate if

$$\nabla \lambda_i(u) \cdot r_i(u) = 0 \qquad \forall u \in \Omega.$$

If the i-th characteristic field is genuinely nonlinear, then we assume that $\nabla \lambda_i(u) \cdot r_i(u) > 0$ for every $u \in \Omega$.

The Riemann problem (A.25) produces n different waves in general, one for each characteristic field. In the case a characteristic field is genuinely nonlinear, then, a shock or a rarefaction wave is produced (as in the scalar case with strictly convex or concave flux). In the case a characteristic field is linearly degenerate, then, a contact discontinuity wave is produced (similar to the transport scalar equations).

There are three possible key cases.

1. **Centered rarefaction waves**. For $u^- \in \Omega$, $i \in \{1, \cdots, n\}$ and $\sigma > 0$, define $R_i(\sigma)(u^-)$ as the solution to

$$\begin{cases} \dot{u} = r_i(u), \\ u(0) = u^-. \end{cases} \tag{A.30}$$

Let $\bar{\sigma} > 0$ and assume that $u^+ = R_i(\bar{\sigma})(u^-)$ for some $i \in \{1, \cdots, n\}$. If the i-th characteristic field is genuinely nonlinear, then the function

$$u(t, x) := \begin{cases} u^-, & \text{if } x < \lambda_i(u^-)t, \\ R_i(\sigma)(u^-), & \text{if } x = \lambda_i(R_i(\sigma)(u^-))t, \ \sigma \in [0, \bar{\sigma}], \\ u^+, & \text{if } x > \lambda_i(u^+)t \end{cases} \qquad (A.31)$$

is an entropy admissible solution to the Riemann problem (A.25). The function $u(t, x)$ in (A.31) is called a centered rarefaction wave.

Remark 19 Note that the construction of u in (A.31) can be done only if $\lambda_i(u^-) < \lambda_i(u^+)$. This condition holds only in the case $\bar{\sigma} > 0$.

2. **Shock waves.** Fix $u^- \in \Omega$ and $i \in \{1, \cdots, n\}$. For some $\sigma_0 > 0$, there exist smooth functions $S_i(u_-) = S_i : [-\sigma_0, \sigma_0] \to \Omega$ and $\lambda_i : [-\sigma_0, \sigma_0] \to \mathbb{R}$ such that:

(a) $f(S_i(\sigma)) - f(u^-) = \lambda_i(\sigma)(S_i(\sigma) - u^-)$ for every $\sigma \in [-\sigma_0, \sigma_0]$
(b) $S_i(0) = u^-, \lambda_i(0) = \lambda_i(u^-)$
(c) $\frac{dS_i(\sigma)}{d\sigma}|_{\sigma=0} = r_i(u^-)$

Let $\bar{\sigma} < 0$ and define $u^+ = S_i(\bar{\sigma})$. If the i-th characteristic field is genuinely nonlinear, then the function

$$u(t, x) := \begin{cases} u^-, & \text{if } x < \lambda_i(\bar{\sigma})t, \\ u^+, & \text{if } x > \lambda_i(\bar{\sigma})t \end{cases} \qquad (A.32)$$

is an entropy admissible solution to the Riemann problem (A.25). The function $u(t, x)$ in (A.32) is called a shock wave.

Remark 20 If $\bar{\sigma} > 0$, then (A.32) is still a weak solution to (A.25), but it does not satisfy the entropy condition.

3. **Contact discontinuities.** Fix $u^- \in \Omega$, $i \in \{1, \cdots, n\}$ and $\bar{\sigma} \in [-\sigma_0, \sigma_0]$. Define $u^+ = S_i(\bar{\sigma})$. If the i-th characteristic field is linearly degenerate, then the function

$$u(t, x) := \begin{cases} u^-, & \text{if } x < \lambda_i(u^-)t, \\ u^+, & \text{if } x > \lambda_i(u^-)t \end{cases} \qquad (A.33)$$

is an entropy admissible solution to the Riemann problem (A.25). The function $u(t, x)$ in (A.33) is called a contact discontinuity.

Definition 46 The waves defined in (A.31), (A.32), and (A.33) are called waves of the i-th family.

For each $\sigma \in \mathbb{R}$ and $i \in \{1, \ldots, n\}$, let us consider the Lax curve

$$\psi_i(\sigma)(\bar{u}) := \begin{cases} R_i(\sigma)(\bar{u}), & \text{if } \sigma \geq 0, \\ S_i(\sigma)(\bar{u}), & \text{if } \sigma < 0, \end{cases} \qquad (A.34)$$

where $\bar{u} \in \Omega$. The value σ is called the strength of the wave of the i-th family, connecting $\bar{u}$ to $\psi_i(\sigma)(\bar{u})$. Moreover, let us consider the composite function

$$\Psi(\sigma_1, \ldots, \sigma_n)(u^-) := \psi_n(\sigma_n) \circ \cdots \circ \psi_1(\sigma_1)(u^-), \qquad (A.35)$$

where $u^- \in \Omega$ and $(\sigma_1, \ldots, \sigma_n)$ belongs to a neighborhood of 0 in $\mathbb{R}^n$. The following result about the local existence of solution to a Riemann problem holds.

Theorem 41 *For every compact set $K \subset \Omega$, there exists $\delta > 0$ such that, for every $u^- \in K$ and for every $u^+ \in \Omega$ with $|u^+ - u^-| \leq \delta$, there exists a unique $(\sigma_1, \ldots, \sigma_n)$ in a neighborhood of $0 \in \mathbb{R}^n$ satisfying*

$$\Psi(\sigma_1, \ldots, \sigma_n)(u^-) = u^+.$$

Moreover, the Riemann problem connecting u^- with u^+ admits an entropy admissible solution, constructed by piecing together the solutions of n Riemann problems.

In the following examples, we consider the Saint-Venant equations, the Aw–Rascle–Zhang model, and the p-system, and we describe the various waves for such systems.

Example 14 The **Saint-Venant** or **shallow water** equations are

$$\begin{cases} \partial_t H + \partial_x (HV) = 0, \\ \partial_t V + \partial_x \left(\frac{V^2}{2} + gH \right) = 0, \end{cases} \qquad (A.36)$$

where $H = H(t, x)$ denotes the water level at time t and at position x, $V = V(t, x)$ is the water velocity, and g is the gravitation constant; see Example 8.

The Jacobian matrix of the flux f has eigenvalues

$$\lambda_1 = V - \sqrt{gH}, \qquad \lambda_2 = V + \sqrt{gH}$$

and right eigenvectors

$$r_1 = \begin{pmatrix} -\dfrac{H}{\sqrt{gH}} \\ 1 \end{pmatrix}, \qquad r_2 = \begin{pmatrix} \dfrac{H}{\sqrt{gH}} \\ 1 \end{pmatrix}.$$

This implies that the system (A.36) is strictly hyperbolic, provided $H > 0$. Moreover,

$$\nabla\lambda_1 \cdot r_1 = \frac{3}{2}, \qquad \nabla\lambda_2 \cdot r_2 = \frac{3}{2},$$

and so the characteristic fields are both genuinely nonlinear.

Let us consider the Riemann problem for the Saint-Venant equations with initial data

$$u(0, x) = \begin{cases} u^- = (H^-, V^-), & \text{if } x < 0, \\ u^+ = (H^+, V^+), & \text{if } x > 0. \end{cases}$$

The equation $\dot{u} = r_i(u)$ gives the following rarefaction curves starting at u^-:

$$R_1 = \left\{ (H, V) : V = -2\sqrt{gH} + 2\sqrt{gH^-} + V^-, \quad H \leq H^- \right\},$$

$$R_2 = \left\{ (H, V) : V = 2\sqrt{gH} - 2\sqrt{gH^-} + V^-, \quad H \geq H^- \right\}.$$

The Rankine–Hugoniot condition, instead, gives the shock curves S_1 and S_2 starting at u^-, which are

$$S_1 = \left\{ (H, V) : V = V^- + 2\sqrt{2g} \, \frac{H^-}{\sqrt{H + H^-}} - \sqrt{2g}\sqrt{H + H^-}, \quad H \geq H^- \right\},$$

$$S_2 = \left\{ (H, V) : V = V^- - 2\sqrt{2g} \, \frac{H^-}{\sqrt{H + H^-}} + \sqrt{2g}\sqrt{H + H^-}, \quad H \leq H^- \right\}.$$

The plots of the Lax curves are in Fig. A.11. The curves R_1, R_2, S_1, and S_2 divide the (H, V) semi-plane ($H > 0$) into four regions A_1, A_2, A_3, and A_4; see Fig. A.11.

If u^+ belongs to one of these curves, then the Riemann problem is solved by a single wave. If instead u^+ is sufficiently near to u^- and belongs to one of the regions A_i, then the solution to the Riemann problem is given by two centered waves. More precisely, if $u^+ \in A_1$, then the solution is given by a rarefaction wave of the first family and by a shock wave of the second family. If $u^+ \in A_2$, then the solution is given by two rarefaction waves. If $u^+ \in A_3$, then the solution is given by two shock waves. If $u^+ \in A_4$, then the solution is given by a shock wave of the first family and by a rarefaction wave of the second family.

Example 15 The **Aw–Rascle–Zhang** model for traffic in conservation form, see Example 6, is

$$\begin{cases} \partial_t \rho + \partial_x (y - \rho p(\rho)) = 0, \\ \partial_t y + \partial_x \left(\frac{y}{\rho}(y - \rho p(\rho)) \right) = 0, \end{cases}$$

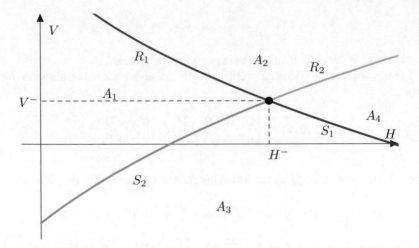

Fig. A.11 The rarefaction and shock curves for the Saint-Venant equations starting from the point (H^-, V^-)

where $\rho = \rho(t, x)$ denotes the density of cars at time t and at position x, $p = p(\rho)$ is a pressure term, $y = \rho(v + p(\rho))$ is a generalized momentum, and $v = v(t, x)$ is the average velocity of cars. In this example, we assume $p(\rho) = \rho^\gamma$, for some $\gamma > 0$.

The Jacobian matrix of the flux f has eigenvalues

$$\lambda_1 = \frac{y}{\rho} - (\gamma + 1)\rho^\gamma, \qquad \lambda_2 = \frac{y}{\rho} - \rho^\gamma$$

and right eigenvectors

$$r_1 = \begin{pmatrix} -\rho \\ -y \end{pmatrix}, \qquad r_2 = \begin{pmatrix} \rho \\ y + \gamma\rho^{\gamma+1} \end{pmatrix}.$$

Note that, since

$$\nabla\lambda_1 \cdot r_1 = \gamma(\gamma + 1)\rho^\gamma, \qquad \nabla\lambda_2 \cdot r_2 = 0,$$

the first characteristic field is genuinely nonlinear, while the second one is linearly degenerate.

Let us consider the Riemann problem for the Aw–Rascle–Zhang model with initial data

$$u(0, x) = \begin{cases} u^- = (\rho^-, y^-), & \text{if } x < 0, \\ u^+ = (\rho^+, y^+), & \text{if } x > 0. \end{cases}$$

Fig. A.12 The rarefaction and shock curves for the Aw–Rascle–Zhang model starting from the point (ρ^-, y^-)

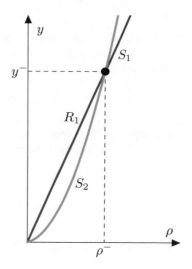

The equation $\dot{u} = r_1(u)$ gives the following rarefaction curve starting at u^-:

$$R_1 = \left\{ (\rho, y) : y = \frac{y^-}{\rho^-} \rho, \quad \rho \le \rho^- \right\}.$$

The Rankine–Hugoniot condition, instead, gives the shock curve S_1 and the contact discontinuity curve S_2 starting at u^-:

$$S_1 = \left\{ (\rho, y) : y = \frac{y^-}{\rho^-} \rho, \quad \rho \ge \rho^- \right\},$$

$$S_2 = \left\{ (\rho, y) : y = \frac{y^-}{\rho^-} \rho + \rho^{\gamma+1} - \rho \left(\rho^-\right)^\gamma, \quad \rho \ge 0 \right\}.$$

The plots of the Lax curves are in Fig. A.12. If u^+ belongs to one of these curves, then the Riemann problem is solved by a single wave. If instead u^+ is sufficiently near to u^- and does not belong to the Lax curves through u^-, then the solution to the Riemann problem is given by two centered waves, one of the first family (rarefaction or shock wave) and one contact discontinuity.

Example 16 The *p*-**system**, see Example 7, is given by

$$\begin{cases} \partial_t \rho + \partial_x q = 0, \\ \partial_t q + \partial_x \left(\frac{q^2}{\rho} + p(\rho) \right) = 0, \end{cases} \tag{A.37}$$

where $\rho > 0$ is the density of the gas and q is the linear momentum density, i.e., $q = \rho v$, where v is the speed of the gas. The function p is the pressure and depends only on the density ρ. Assume that p is of class C^2 and

$$p(\rho) > 0, \quad p'(\rho) > 0, \quad p''(\rho) \geq 0$$

for every $\rho > 0$. A typical example is the γ-pressure law $p(\rho) = k\rho^\gamma$ for $k > 0$ and $\gamma \geq 1$. The Jacobian matrix for the flux f is

$$A(U) = \begin{pmatrix} 0 & 1 \\ -\frac{q^2}{\rho^2} + p'(\rho) & 2\frac{q}{\rho} \end{pmatrix},$$

which has the distinct eigenvalues

$$\lambda_1 = \frac{q}{\rho} - \sqrt{p'(\rho)}, \qquad \lambda_2 = \frac{q}{\rho} + \sqrt{p'(\rho)},$$

and the corresponding right eigenfunctions

$$r_1 = \begin{pmatrix} \rho \\ q - \rho\sqrt{p'(\rho)} \end{pmatrix}, \qquad r_2 = \begin{pmatrix} \rho \\ q + \rho\sqrt{p'(\rho)} \end{pmatrix}.$$

This implies that the system (A.37) is strictly hyperbolic. Moreover,

$$\nabla\lambda_1 \cdot r_1 = -\sqrt{p'(\rho)} - \rho\frac{p''(\rho)}{2\sqrt{p'(\rho)}}, \qquad \nabla\lambda_2 \cdot r_2 = \sqrt{p'(\rho)} + \rho\frac{p''(\rho)}{2\sqrt{p'(\rho)}},$$

and so the characteristic fields are both genuinely nonlinear.

Let us consider the Riemann problem for (A.37) with initial data

$$u(0, x) = \begin{cases} u^- = (\rho^-, q^-), & \text{if } x < 0, \\ u^+ = (\rho^+, q^+), & \text{if } x > 0. \end{cases}$$

The equations $\dot{u} = r_i(u)$ give the following rarefaction curves starting at u^-:

$$R_1 = \left\{ (\rho, q) : q = \frac{\rho q^-}{\rho^-} - \rho \int_{\rho^-}^{\rho} \frac{\sqrt{p'(r)}}{r} dr, \quad \rho \leq \rho^- \right\},$$

$$R_2 = \left\{ (\rho, q) : q = \frac{\rho q^-}{\rho^-} + \rho \int_{\rho^-}^{\rho} \frac{\sqrt{p'(r)}}{r} dr, \quad \rho \geq \rho^- \right\}.$$

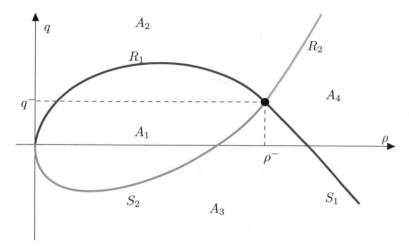

Fig. A.13 The rarefaction and shock curves for the p-system (A.37) starting from the point (ρ^-, q^-)

The Rankine–Hugoniot condition, instead, gives the shock curves S_1 and S_2 starting at u^-:

$$S_1 = \left\{ (\rho, q) : q = \frac{\rho q^-}{\rho^-} - \sqrt{\frac{\rho}{\rho^-}(\rho - \rho^-)\left(p(\rho) - p(\rho^-)\right)}, \quad \rho \geq \rho^- \right\},$$

$$S_2 = \left\{ (\rho, q) : q = \frac{\rho q^-}{\rho^-} - \sqrt{\frac{\rho}{\rho^-}(\rho - \rho^-)\left(p(\rho) - p(\rho^-)\right)}, \quad \rho \leq \rho^- \right\}.$$

The situation is described in Fig. A.13. The curves R_i and S_i divide the (ρ, q)-plane into four regions A_1, A_2, A_3, and A_4. If u^+ belongs to one of these curves, then the Riemann problem is solved by a single wave. If instead u^+ is sufficiently near to u^- and belongs to one of the regions A_i, then the solution to the Riemann problem is given by two centered waves. More precisely, if $u^+ \in A_1$, then the solution is given by a rarefaction wave of the first family and by a shock wave of the second family. If $u^+ \in A_2$, then the solution is given by two rarefaction waves. If $u^+ \in A_3$, then the solution is given by two shock waves. If $u^+ \in A_4$, then the solution is given by a shock wave of the first family and by a rarefaction wave of the second family.

A.7 The Cauchy Problem

Once the solution to Riemann problems is defined, one can construct the solution to a Cauchy problem, by using the wave-front tracking technique.

To this aim, consider the Cauchy problem

$$\begin{cases} \partial_t u + \partial_x f(u) = 0, \\ u(0, \cdot) = u_0(\cdot), \end{cases} \tag{A.38}$$

where $f : \Omega \to \mathbb{R}^n$ is a smooth flux and $u_0 \in L^1(\mathbb{R}; \Omega) \cap BV(\mathbb{R}; \Omega)$.

We start considering the scalar case, while the system case, much more delicate, will be only sketched.

A.7.1   Wave-Front Tracking for the Scalar Case

Assume that the flux function $f : \mathbb{R} \to \mathbb{R}$ is smooth and strictly convex or concave function. Choose a sequence of piecewise constant functions $\{u_0^\nu\}_\nu$ satisfying

$$\mathrm{TV}(u_0^\nu) \le \mathrm{TV}(u_0), \tag{A.39}$$

$$\|u_0^\nu\|_{\mathbf{L}^\infty} \le \|u_0\|_{\mathbf{L}^\infty}, \tag{A.40}$$

$$\|u_0^\nu - u_0\|_{\mathbf{L}^1} < \frac{1}{\nu}, \tag{A.41}$$

for every $\nu \in \mathbb{N}$; see Fig. A.14. This is possible since $u_0 \in \mathbf{L}^1(\mathbb{R}; \mathbb{R}) \cap \mathbf{BV}(\mathbb{R}; \mathbb{R})$; see [51, Lemma 2.2].

Fix $\nu \in \mathbb{N}$. By (A.39), u_0^ν has a finite number of discontinuities, say $x_1 < \cdots < x_N$. For each $i = 1, \cdots, N$, we approximately solve the Riemann problem generated by the jump $(u_0^\nu(x_i-), u_0^\nu(x_i+))$ with piecewise constant functions of the type $\psi(\frac{x-x_i}{t})$, where $\psi : \mathbb{R} \to \mathbb{R}$. More precisely, if the Riemann problem generated by $(u_0^\nu(x_i-), u_0^\nu(x_i+))$ admits an exact solution containing a shock, then

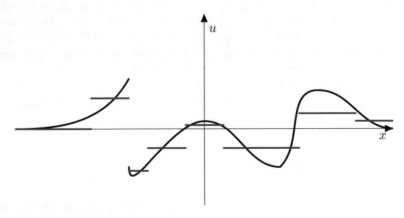

Fig. A.14 A piecewise constant approximation (in blue) of the initial datum $\bar{u}$ satisfying (A.40) and (A.41)

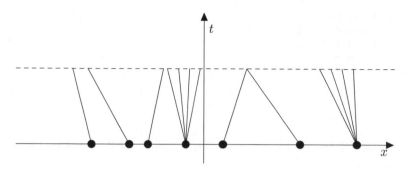

Fig. A.15 The wave-front tracking construction until the first time of interaction

$\psi(\frac{x-x_i}{t})$ is the exact solution, while if a rarefaction wave appears, then we split it in a centered rarefaction fan, containing a sequence of jumps of size at most $\frac{1}{\nu}$, travelling with a speed between the characteristic speeds of the states connected. In this way, we are able to construct an approximate solution $u^\nu(t, x)$ until a time t_1, where at least two wave-fronts interact together; see Fig. A.15.

Remark 21 Notice that it is possible to avoid that three of more wave-fronts interact at the same time slightly changing the speed of some wave-fronts. This may introduce a small error in the approximate solution with respect to the exact one.

At time $t = t_1, u^\nu(t_1, \cdot)$ is clearly a piecewise constant function. So we can repeat the previous construction until a second interaction time $t = t_2$ and so on. In order to prove that a wave-front tracking approximate solution exists for every $t > 0$, we need to estimate

1. The number of waves
2. The number of interactions between waves
3. The total variation of the approximate solution

The first two estimates are concerned with the possibility to construct a piecewise constant approximate solution. The third estimate, instead, is related to the convergence of the approximate solutions toward an exact solution.

Remark 22 The two first bounds are nontrivial for the system case, and it is necessary to introduce simplified solutions to Riemann problems and/or non-physical waves.

The next lemma shows that the number of interactions is finite.

Lemma 13 *The number of wave-fronts for the approximate solution u^ν is not increasing with respect to the time and so u^ν is defined for every $t \geq 0$. Moreover, the number of interactions between waves is bounded by the number of wave-fronts.*

Fig. A.16 Interaction
between two wave-fronts. The
first one connects the states u_l
and u_m, while the second one
connects u_m with u_r

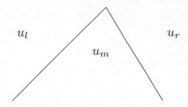

Proof Consider two wave-fronts interacting together. The wave fronts can be:

1. Two shocks
2. Two rarefaction fronts
3. A shock and a rarefaction front

The speeds of waves imply that the case of two rarefaction fronts cannot happen. In fact, suppose that two rarefaction shocks interact together at a certain time. Denote with u_l, u_m, and u_r, respectively, the left, middle, and right states as in Fig. A.16. Since these waves are rarefaction shocks, we have

$$f'(u_l) < f'(u_m) < f'(u_r),$$

where $f'(u)$ is the characteristic speed of state u. Therefore, the wave connecting u_l to u_m has a speed less than or equal to the speed of the wave connecting u_m to u_r, and the wave-fronts cannot interact.

So the remaining possibilities are the following:

1. Two shocks. In this case, it is clear that after the interaction, a single shock wave is created. So the number of waves decreases by 1.
2. A shock and a rarefaction front. In this case, either a single shock wave is produced as in the previous possibility, or a single rarefaction shock is created. In fact, if the exact solution to the Riemann problem at the interaction time is given by a rarefaction wave, then the size of the rarefaction wave is less than or equal to the size of the rarefaction front, which is less than or equal to $1/\nu$. This implies that the wave is split in a single rarefaction shock. Thus the number of waves decreases by 1.

Therefore, we conclude that at each interaction the number of wave-fronts decreases at least by 1 and so the lemma is proved.   □

Lemma 14 *The total variation of $u^\nu(t, \cdot)$ is not increasing with respect to time. Therefore, for each $t \geq 0$,*

$$\text{TV}\,(u^\nu(t, \cdot)) \leq \text{TV}\,(u_0). \tag{A.42}$$

Proof It is clear that the total variation may vary only at interaction times.

Consider an interaction of two wave-fronts at time $\bar{t}$. Let us call by u_l, u_m, and u_r, respectively, the left, the middle, and the right states of the wave-fronts; see Fig. A.16.

The interaction between the two waves produces a single wave connecting u_l with u_r. The variation before $t = \bar{t}$ due to the interacting waves is given by $|u_l - u_m| + |u_m - u_r|$, while the variation after $t = \bar{t}$ due to the wave produced is given by $|u_l - u_r|$. The triangular inequality implies that

$$|u_l - u_r| \leq |u_l - u_m| + |u_m - u_r|$$

and so the proof is finished. $\qquad\square$

The following theorem holds.

Theorem 42 *Let $f : \mathbb{R} \to \mathbb{R}$ be a smooth strictly convex or strictly concave function and $u_0 \in \mathbf{L}^1(\mathbb{R}; \mathbb{R}) \cap \mathbf{BV}(\mathbb{R}; \mathbb{R})$. Then there exists an entropy admissible solution $u(t, x)$ to the Cauchy problem (A.38), defined for every $t \geq 0$. Moreover,*

$$\|u(t, \cdot)\|_{\mathbf{L}^\infty} \leq \|u_0(\cdot)\|_{\mathbf{L}^\infty} \tag{A.43}$$

for every $t \geq 0$.

Proof For every $v \in \mathbb{N}$, construct a wave-front tracking approximate solution u^v as before in this section.

Clearly, we have

$$\left|u^v(t, x)\right| \leq \left|u^v(0, x)\right| \leq \|u_0\|_{\mathbf{L}^\infty} \tag{A.44}$$

for every $v \in \mathbb{N}$, $t \geq 0$, and $x \in \mathbb{R}$. By Lemma 14,

$$\mathrm{TV}\left(u^v(t, \cdot)\right) \leq \mathrm{TV}\left(u_0\right), \tag{A.45}$$

for every $t \geq 0$ and $v \in \mathbb{N}$. Finally, the maps $t \mapsto u^v(t, \cdot)$ are uniformly Lipschitz continuous with values in $\mathbf{L}^1(\mathbb{R}; \mathbb{R})$. Therefore, by Helly's theorem (see, for example, [51, Theorem 2.4]), we can extract a subsequence, denoted again by $u^v(t, x)$, converging to some function $u(t, x)$ in $\mathbf{L}^1([0, +\infty[\times\mathbb{R}; \mathbb{R})$. Since $\|u^v(0, \cdot) - u_0(\cdot)\|_{\mathbf{L}^1} \to 0$, then the initial condition clearly holds.

It remains to prove that $u(t, x)$ is a weak solution to the Cauchy problem (A.38) and that it is entropy admissible. To prove the first claim, fix $T > 0$ and an arbitrary $\mathbf{C}^1$ function ψ with compact support in $]-\infty, T[\times\mathbb{R}$. We need to prove that

$$\int_0^T \int_{\mathbb{R}} (u\partial_t \psi + f(u)\psi) \, \mathrm{d}x \, \mathrm{d}t + \int_{\mathbb{R}} u_0(x)\psi(0, x) \, \mathrm{d}x = 0.$$

It is sufficient to prove that

$$\lim_{\nu \to +\infty} \left[\int_0^T \!\!\! \int_{\mathbb{R}} \left(u^\nu \partial_t \psi + f(u^\nu) \partial_x \psi \right) dx \, dt + \int_{\mathbb{R}} u^\nu(0, x) \psi(0, x) \, dx \right] = 0. \quad (A.46)$$

Fix $\nu \in \mathbb{N}$. At every $t \in [0, T]$, call $x_1(t) < \cdots < x_N(t)$ the points where $u^\nu(t, \cdot)$ has a jump and set

$$\Delta u^\nu(t, x_\alpha) := u^\nu(t, x_\alpha +) - u^\nu(t, x_\alpha -),$$
$$\Delta f(u^\nu(t, x_\alpha)) := f(u^\nu(t, x_\alpha +)) - f(u^\nu(t, x_\alpha -)).$$

The lines $x_\alpha(t)$ divide $[0, T] \times \mathbb{R}$ into a finite number of regions, say Γ_j, where u^ν is constant. Applying the Divergence Theorem to the vector field $(\psi u^\nu, \psi f(u^\nu))$ and splitting the integral (A.46) over the regions Γ_j, we obtain that the integral (A.46) can be rewritten in the form

$$\int_0^T \sum_\alpha \left[\dot{x}_\alpha(t) \Delta u^\nu(t, x_\alpha) - \Delta f(u^\nu(t, x_\alpha)) \right] \psi(t, x_\alpha(t)) \, dt. \quad (A.47)$$

If x_α is a shock wave, then

$$\dot{x}_\alpha(t) \cdot \Delta u^\nu(t, x_\alpha) - \Delta f(u^\nu(t, x_\alpha)) = 0,$$

while if x_α is a rarefaction wave, then

$$\dot{x}_\alpha(t) \cdot \Delta u^\nu(t, x_\alpha) - \Delta f(u^\nu(t, x_\alpha))$$

depends linearly on the $\mathbf{L}^\infty$ distance between $u^\nu(0, \cdot)$ and u_0. Splitting the summation in (A.47) over waves of the same type, we deduce that the previous integral tends to 0 as $\nu \to +\infty$, concluding that $u(t, x)$ is a weak solution to the Cauchy problem.

Fix now η a convex entropy with a corresponding entropy flux q. It remains to prove that

$$\liminf_{\nu \to +\infty} \int_0^T \int_{\mathbb{R}} \left[\eta(u^\nu) \partial_t \psi + q(u^\nu) \partial_x \psi \right] dx \, dt \geq 0$$

for every $\mathbf{C}^1$ positive function ψ with compact support. Using again the Divergence Theorem as before, we need to prove that

$$\liminf_{\nu \to +\infty} \int_0^T \sum_\alpha \left[\dot{x}_\alpha(t) \Delta \eta(u^\nu(t, x_\alpha)) - \Delta q(u^\nu(t, x_\alpha)) \right] \psi(t, x_\alpha) \, dt \geq 0,$$

where

$$\Delta\eta(u^\nu(t, x_\alpha)) := \eta(u^\nu(t, x_\alpha+)) - \eta(u^\nu(t, x_\alpha-)),$$

$$\Delta q(u^\nu(t, x_\alpha)) := q(u^\nu(t, x_\alpha+)) - q(u^\nu(t, x_\alpha-)).$$

Using the same estimates as in the previous case, we conclude.   □

A.7.2   The System Case

For systems, the construction of wave-front tracking approximations is more complex, because more types of interactions may happen. In particular, the bounds on the number of waves, interactions, and **BV** norms are no more directly obtained.

Let us start giving some total variation estimates for interaction of waves along a wave-front tracking approximation. These permit to illustrate the ideas for obtaining the needed bounds in the system case. The constants in the estimates depend on the total variation of the initial data, which is assumed to be sufficiently small.

Consider a wave of the i-th family of strength σ_i interacting with a wave of the j-th family of strength σ_j, $i \neq j$, and indicate by σ_k' ($k \in \{1, \ldots, n\}$) the strengths of the new waves produced by the interaction. Then, it holds

$$\left|\sigma_i - \sigma_i'\right| + \left|\sigma_j - \sigma_j'\right| + \sum_{k\neq i, j} \left|\sigma_k'\right| \leq C \left|\sigma_i \sigma_j\right|. \tag{A.48}$$

For the case $i = j$, let us indicate by $\sigma_{i,1}$ and $\sigma_{i,2}$ the strengths of the interacting waves, and then it holds

$$\left|\sigma_{i,1} + \sigma_{i,2} - \sigma_i'\right| + \sum_{k\neq i} \left|\sigma_k'\right| \leq C \left|\sigma_{i,1}\sigma_{i,2}\right|. \tag{A.49}$$

One can now fix a parameter δ_ν and split rarefactions in rarefaction fans with fronts of strength at most δ_ν. Also, at each interaction time, one solves exactly the new Riemann problem, eventually splitting the rarefaction waves in rarefaction fans, only if the product of interacting waves is bigger than δ_ν. Otherwise, one solves the Riemann problem only with waves of the families of the interacting ones, the error being transported along a *non-physical wave*, traveling at a speed bigger than all waves. In this way, it is possible to control the number of waves and interactions, and then let δ_ν go to zero. For details, see [51, Lemma 7.2].

Consider now a wave-front tracking approximate solution u^ν, and let $x_\alpha(t)$, of family i_α and strength σ_α, indicate the discontinuities of $u^\nu(t)$. We say that two discontinuities are interacting if $x_\alpha < x_\beta$ and either $i_\alpha > i_\beta$ or $i_\alpha = i_\beta$ and at least one of the two waves is a shock. We define the Glimm functional computed at $u^\nu(t)$ as

$$Y(u^\nu(t)) = \mathrm{TV}(u^\nu(t)) + C_1 \, Q(u^\nu(t)),$$

where C_1 is a constant to be chosen suitably and

$$Q(u^\nu(t)) = \sum |\sigma_\alpha \sigma_\beta|,$$

where the sum is over interacting waves. One can easily prove that the functional Y is equivalent to the functional measuring the total variation. Clearly, such functional changes only at interaction times. Using the interaction estimates (A.48) and (A.49), at an interaction time $\bar{t}$, we get

$$\left|\mathrm{TV}(u^\nu(\bar{t}+)) - \mathrm{TV}(u^\nu(\bar{t}-))\right| \le C \left|\sigma_i \sigma_j\right|,$$

$$Q(u^\nu(\bar{t}+)) - Q(u^\nu(\bar{t}-)) \le -C_1 \left|\sigma_i \sigma_j\right| + C \left|\sigma_i \sigma_j\right| \mathrm{TV}(u^\nu(\bar{t}-)).$$

Therefore,

$$Y(u^\nu(\bar{t}+)) - Y(u^\nu(\bar{t}-)) \le \left|\sigma_i \sigma_j\right| \left[C - C_1 + C\mathrm{TV}(u^\nu(\bar{t}-))\right].$$

On the other side, for every t,

$$\mathrm{TV}(u^\nu(t)) \le Y(u^\nu(t)).$$

Then, choosing $C_1 > C$ and assuming that $\mathrm{TV}(u^\nu(0))$ is sufficiently small, one has that Y is decreasing along a wave-front tracking approximate solution and so the total variation is controlled.

A.8   Boundary Conditions for Scalar Conservation Laws

Here we describe briefly the problem of boundary conditions for conservation laws. We focus the attention on the following problem:

$$\begin{cases} \partial_t u + \partial_x f(u) = 0, & t > 0,\ x \in (a, b), \\ u(t, a) = u_a(t), & t > 0, \\ u(t, b) = u_b(t), & t > 0, \\ u(0, x) = u_0(x), & x \in (a, b), \end{cases} \tag{A.50}$$

where $a < b$, the unknown u is defined on $[0, +\infty[\times]a, b[$ with values on $\mathbb{R}$, the flux $f : \mathbb{R} \to \mathbb{R}$ is a smooth function, u_a and u_b are the left and right boundary conditions, and u_0 is the initial data. The case with a single boundary can be treated in a similar way. First note that boundary conditions cannot be always interpreted in the classical sense. The next examples show different roles played by boundary conditions.

Example 17 Consider the following boundary value problem for the transport equation:

$$\begin{cases} \partial_t u - \partial_x u = 0, & t > 0,\ x > 0, \\ u(t, 0) = 1, & t > 0, \\ u(0, x) = 0, & x > 0. \end{cases} \tag{A.51}$$

By the method of characteristics, the solution to the conservation law $\partial_t u - \partial_x u = 0$, with the initial condition $u(0, \cdot) = 0$, is the function $u(t, x) = 0$ for every $t \geq 0$ and $x \geq 0$. In this case, the boundary condition $u(t, 0) = 1$ in (A.51) cannot be attained, and it plays no role in the construction of the solution.

Example 18 Consider the following boundary value problem for the transport equation:

$$\begin{cases} \partial_t u + \partial_x u = 0, & t > 0,\ x > 0, \\ u(t, 0) = 1, & t > 0, \\ u(0, x) = 0, & x > 0. \end{cases} \tag{A.52}$$

By the method of characteristics, the solution to the problem (A.52) is

$$u(t, x) = \begin{cases} 1, & \text{if } t > 0,\ 0 \leq x < t, \\ 0, & \text{otherwise.} \end{cases}$$

In this case, the boundary condition $u(t, 0) = 1$ in (A.52) is attained for every $t > 0$.

Boundary conditions for conservation laws have to be interpreted in a weak sense. Various formulations are present in the literature; see, for example, [29, 123, 212]. Following [29], we give the following definition of solution to (A.50).

Definition 47 A function

$$u : C^0 \left([0, +\infty[; L^1 \left( ]a, b[; \mathbb{R} \right) \right)$$

such that $u(t)$ has finite total variation for every $t \geq 0$ is a weak entropy admissible solution to the initial-boundary value problem (A.50) if the following conditions hold:

1. For every C^1 function $\psi \geq 0$ with compact support in $]0, +\infty[\times]a, b[$ and for every entropy–entropy flux pair (η, q), it holds

$$\int_0^{+\infty} \int_a^b (\eta(u)\partial_t \psi + q(u)\partial_x \psi)\ dx\ dt \geq 0. \tag{A.53}$$

2. The limit

$$\lim_{t \to 0^+} \|u(t, \cdot) - u_0(\cdot)\|_{\mathbf{L}^1(]a,b[)} = 0 \qquad (A.54)$$

holds.

3. The weak boundary condition at $x = a$

$$\max_{k \in [\alpha(t),\beta(t)]} \operatorname{sgn}(u(t, a+) - u_a(t)) [f(u(t, a+)) - f(k)] = 0 \qquad (A.55)$$

holds for a.e. $t > 0$, where

$$\alpha(t) = \min\{u(t, a+), u_a(t)\}, \qquad \beta(t) = \max\{u(t, a+), u_a(t)\},$$

and $u(t, a+)$ denotes the right trace at $x = a$ of $u(t, \cdot)$.

4. The weak boundary condition at $x = b$

$$\min_{k \in [\gamma(t),\delta(t)]} \operatorname{sgn}(u(t, b-) - u_b(t)) [f(u(t, b-)) - f(k)] = 0 \qquad (A.56)$$

holds for a.e. $t > 0$, where

$$\gamma(t) = \min\{u(t, b-), u_b(t)\}, \qquad \delta(t) = \max\{u(t, b-), u_b(t)\},$$

and $u(t, b-)$ denotes the left trace at $x = b$ of $u(t, \cdot)$.

A.8.1   The Left Boundary Condition for the Riemann Problem

We describe in detail the solution to the following boundary value: problem

$$\begin{cases} \partial_t u + \partial_x f(u) = 0, & t > 0, \ x > 0, \\ u(t, 0) = \tilde{u}, & t > 0, \\ u(0, x) = u_0, & x > 0, \end{cases} \qquad (A.57)$$

where $u \in [0, 1]$, $\tilde{u}, u_0 \in [0, 1]$ are constants, and $f : [0, 1] \to \mathbb{R}$ is a $\mathbf{C}^2$ strictly concave function satisfying $f(0) = f(1) = 0$. Denote by $\sigma \in]0, 1[$ the point of maximum for f and by $\bar{u} \in [0, 1]$ the right trace at $x = 0$ of a solution to (A.57) according to Definition 47. We finally denote

$$\alpha = \min\{\bar{u}, \tilde{u}\}, \qquad \beta = \max\{\bar{u}, \tilde{u}\}.$$

Since $\bar{u}$ denotes the right trace at $x = 0$ of a solution to (A.57), we may suppose that the classical Riemann problem

$$\begin{cases} \partial_t u + \partial_x f(u) = 0, & t > 0, \ x \in \mathbb{R}, \\ u(0,x) = \bar{u}, & x < 0, \\ u(0,x) = u_0, & x > 0 \end{cases}$$

is solved with waves with non-negative speed, i.e.,

$$\begin{aligned} &\text{either} && \bar{u} = u_0, \\ &\text{or} && \bar{u} < u_0 \text{ and } f(\bar{u}) < f(u_0), \\ &\text{or} && u_0 < \bar{u} \le \sigma. \end{aligned}$$

We deduce now the relations between the trace $\bar{u}$ and the boundary datum $\tilde{u}$, so that condition (A.55) is satisfied. Assume first that $\tilde{u} < \sigma$. We have different possibilities.

1. $\bar{u} < \tilde{u}$. In this case, we deduce that

$$\max_{k \in [\alpha,\beta]} \text{sgn}\,(\bar{u} - \tilde{u})\,[f(\bar{u}) - f(k)] = \max_{k \in [\bar{u},\tilde{u}]} [f(k) - f(\bar{u})]$$

$$= f(\tilde{u}) - f(\bar{u}) > 0.$$

So condition (A.55) is not satisfied, i.e., this case does not happen.
2. $\bar{u} = \tilde{u}$. In this case, we deduce that

$$\max_{k \in [\alpha,\beta]} \text{sgn}\,(\bar{u} - \tilde{u})\,[f(\bar{u}) - f(k)] = \max_{k \in [\bar{u},\tilde{u}]} [f(k) - f(\bar{u})] = 0.$$

Hence condition (A.55) holds.
3. $\tilde{u} < \bar{u} \le \sigma$. In this case, we deduce that

$$\max_{k \in [\alpha,\beta]} \text{sgn}\,(\bar{u} - \tilde{u})\,[f(\bar{u}) - f(k)] = \max_{k \in [\tilde{u},\bar{u}]} [f(\bar{u}) - f(k)]$$

$$= f(\bar{u}) - f(\tilde{u}) > 0.$$

So condition (A.55) is not satisfied, i.e., this case does not happen.
4. $\tilde{u} < \sigma < \bar{u}$. In this case, we deduce that

$$\max_{k \in [\alpha,\beta]} \text{sgn}\,(\bar{u} - \tilde{u})\,[f(\bar{u}) - f(k)] = \max_{k \in [\tilde{u},\bar{u}]} [f(\bar{u}) - f(k)]$$

$$= \begin{cases} f(\bar{u}) - f(\tilde{u}) > 0, & \text{if } f(\tilde{u}) < f(\bar{u}), \\ 0, & \text{otherwise.} \end{cases}$$

Hence condition (A.55) is satisfied only if $f(\tilde{u}) \ge f(\bar{u})$.

Consider now the case $\tilde{u} \ge \sigma$. We have different possibilities.

1. $\bar{u} < \sigma \le \tilde{u}$. In this case, we deduce that

$$\max_{k\in[\alpha,\beta]} \text{sgn}\,(\bar{u} - \tilde{u})\,[f\,(\bar{u}) - f(k)] = \max_{k\in[\bar{u},\tilde{u}]} [f(k) - f\,(\bar{u})]$$

$$= f(\sigma) - f(\bar{u}) > 0.$$

So condition (A.55) is not satisfied, i.e., this case does not happen.

2. $\sigma \le \bar{u} < \tilde{u}$. In this case, we deduce that

$$\max_{k\in[\alpha,\beta]} \text{sgn}\,(\bar{u} - \tilde{u})\,[f\,(\bar{u}) - f(k)] = \max_{k\in[\bar{u},\tilde{u}]} [f(k) - f\,(\bar{u})] = 0.$$

Hence condition (A.55) is satisfied.

3. $\bar{u} = \tilde{u}$. In this case, we deduce that

$$\max_{k\in[\alpha,\beta]} \text{sgn}\,(\bar{u} - \tilde{u})\,[f\,(\bar{u}) - f(k)] = 0.$$

Hence condition (A.55) is satisfied.

4. $\tilde{u} < \bar{u}$. In this case, we deduce that

Table A.1 Relations between the boundary datum $\tilde{u}$ and the admissible traces $\bar{u}$ of a solution to (A.57) satisfying condition (A.55) of Definition 47

Boundary datum $\tilde{u}$	Traces $\bar{u}$ satisfying (A.55)
$\tilde{u} < \sigma$	either $\bar{u} = \tilde{u}$ or $\bar{u} > \sigma$ and $f(\bar{u}) \le f(\tilde{u})$
$\tilde{u} \ge \sigma$	$\bar{u} \ge \sigma$

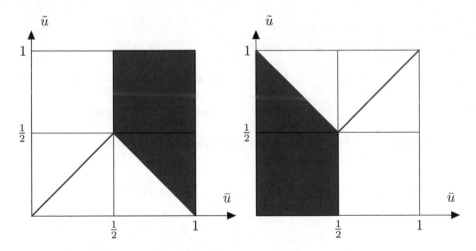

Fig. A.17 Relations between the boundary datum and the trace of a solution in the case of $f(u) = u(1 - u)$. The admissible regions are colored in red. Left: relations described in Table A.1. Right: relations described in Table A.2

$$\max_{k \in [\alpha, \beta]} \operatorname{sgn}(\bar{u} - \tilde{u})\, [f(\bar{u}) - f(k)] = \max_{k \in [\tilde{u}, \bar{u}]} [f(\bar{u}) - f(k)] = 0.$$

Hence condition (A.55) is satisfied.

We summarize all the previous results in Table A.1 and in Fig. A.17, left.

A.8.2   The Right Boundary Condition for the Riemann Problem

Similar to Sect. A.8.1, we briefly describe the solution to the following boundary value problem:

$$\begin{cases} \partial_t u + \partial_x f(u) = 0, & t > 0,\ x < 0, \\ u(t, 0) = \tilde{u}, & t > 0, \\ u(0, x) = u_0, & x < 0, \end{cases} \tag{A.58}$$

where $u \in [0, 1]$, $\tilde{u}, u_0 \in [0, 1]$ are constants, and $f : [0, 1] \to \mathbb{R}$ is a $\mathbf{C}^2$ strictly concave function satisfying $f(0) = f(1) = 0$. Denote by $\sigma \in\,]0, 1[$ the point of maximum for f and by $\bar{u} \in [0, 1]$ the left trace at $x = 0$ of a solution to (A.58), and define $\gamma = \min\{\bar{u}, \tilde{u}\}$, $\delta = \max\{\bar{u}, \tilde{u}\}$.

We deduce now the relations between the trace $\bar{u}$ and the boundary datum $\tilde{u}$, so that condition (A.56) of Definition 47 is satisfied. Assume first that $\tilde{u} \leq \sigma$. We have different possibilities.

1. $\bar{u} \leq \tilde{u}$. In this case, we deduce that

$$\min_{k \in [\gamma, \delta]} \operatorname{sgn}(\bar{u} - \tilde{u})\, [f(\bar{u}) - f(k)] = \min_{k \in [\bar{u}, \tilde{u}]} [f(k) - f(\bar{u})] = 0.$$

Hence condition (A.56) holds.
2. $\tilde{u} < \bar{u} \leq \sigma$. In this case, we deduce that

$$\min_{k \in [\gamma, \delta]} \operatorname{sgn}(\bar{u} - \tilde{u})\, [f(\bar{u}) - f(k)] = \min_{k \in [\tilde{u}, \bar{u}]} [f(\bar{u}) - f(k)] = 0.$$

Hence condition (A.56) holds.
3. $\tilde{u} \leq \sigma < \bar{u}$. In this case, we deduce that

$$\min_{k \in [\gamma, \delta]} \operatorname{sgn}(\bar{u} - \tilde{u})\, [f(\bar{u}) - f(k)] = \min_{k \in [\tilde{u}, \bar{u}]} [f(\bar{u}) - f(k)]$$

$$= f(\bar{u}) - f(\sigma) < 0.$$

Hence condition (A.56) does not hold.

Consider now the case $\tilde{u} > \sigma$. We have different possibilities.

1. $\bar{u} \le \sigma < \tilde{u}$. In this case, we deduce that

$$\min_{k \in [\gamma, \delta]} \mathrm{sgn} \, (\bar{u} - \tilde{u}) \, [f \, (\bar{u}) - f(k)] = \min_{k \in [\bar{u}, \tilde{u}]} [f(k) - f \, (\bar{u})]$$

$$= \begin{cases} 0, & \text{if } f(\bar{u}) \le f(\tilde{u}), \\ f(\tilde{u}) - f(\bar{u}), & \text{if } f(\bar{u}) > f(\tilde{u}). \end{cases}$$

So condition (A.56) is satisfied only if $f(\bar{u}) \le f(\tilde{u})$.

2. $\sigma < \bar{u} < \tilde{u}$. In this case, we deduce that

$$\min_{k \in [\gamma, \delta]} \mathrm{sgn} \, (\bar{u} - \tilde{u}) \, [f \, (\bar{u}) - f(k)] = \min_{k \in [\bar{u}, \tilde{u}]} [f(k) - f \, (\bar{u})]$$

$$= f(\tilde{u}) - f(\bar{u}) < 0.$$

Hence condition (A.56) does not hold.

3. $\bar{u} = \tilde{u}$. In this case, we deduce that

$$\min_{k \in [\gamma, \delta]} \mathrm{sgn} \, (\bar{u} - \tilde{u}) \, [f \, (\bar{u}) - f(k)] = 0.$$

Hence condition (A.56) is satisfied.

4. $\tilde{u} < \bar{u}$. In this case, we deduce that

$$\min_{k \in [\gamma, \delta]} \mathrm{sgn} \, (\bar{u} - \tilde{u}) \, [f \, (\bar{u}) - f(k)] = \min_{k \in [\tilde{u}, \bar{u}]} [f \, (\bar{u}) - f(k)]$$

$$= f(\bar{u}) - f(\tilde{u}) < 0.$$

Hence condition (A.56) is not satisfied.

We summarize all the previous results in Table A.2 and in Fig. A.17, right.

Table A.2 Relations between the boundary datum $\tilde{u}$ and the admissible traces $\bar{u}$ of a solution to (A.57) satisfying condition (A.56) of Definition 47

Boundary datum $\tilde{u}$	Traces $\bar{u}$ satisfying (A.56)
$\tilde{u} \le \sigma$	$\bar{u} \le \sigma$
$\tilde{u} > \sigma$	either $\bar{u} = \tilde{u}$ or $\bar{u} < \sigma$ and $f(\bar{u}) \le f(\tilde{u})$

Appendix B
Models for Vehicular Traffic and Conservation Laws on Networks

In this appendix, we provide various results for hyperbolic conservation laws on networks. We first start with our key example of vehicular traffic and then turn to other possible applications.

The development of models for vehicular traffic on networks started with the pioneering work of Holden and Risebro [178] in 1995. After around a decade, a renewed interest gave rise to many models, see Chitour and Piccoli [70], Coclite et al. [76], Garavello and Piccoli [141, 142], Herty et al. [175, 177], Holden and Risebro [178], and Lebacque and Khoshyaran [207]. The same models were used for other applications such as telecommunication networks (see D'Apice et al. [108]), gas pipelines networks (see Banda et al. [26] and Colombo and Garavello [77]), and supply chains (see Armbruster et al. [20], D'Apice and Manzo [107], and Göttlich et al. [156]).

B.1 Lighthill–Whitham–Richard Model for vehicular Traffic on Networks

The most well-known macroscopic model for traffic flow, the celebrated Lighthill–Whitham–Richard (briefly LWR), consists of a single conservation law. One first starts from the conservation of vehicles written as

$$\frac{d}{dt} \int_{x_1}^{x_2} u(t, x) \, dx = f(t, x_1) - f(t, x_2),$$

where x_1 and x_2 are two locations on a stretch of road, u is the car density, and f represents the flux. In other words, the time variation in the amount of cars between x_1 and x_2 is given by the difference between the incoming flux at x_1 and the outgoing flux at x_2. Recalling that $f = f(u, v) = uv$, where v is the average speed, and

© The Author(s) 2022
A. Bayen et al., *Control Problems for Conservation Laws with Traffic Applications*,
PNLDE Subseries in Control 99, https://doi.org/10.1007/978-3-030-93015-8

differentiating, we get

$$\partial_t u + \partial_x f(u, v) = 0. \tag{B.1}$$

Equation (B.1) does not provide self-contained mathematical model, since it depends on two variables u and v. To overcome this difficulty, the LWR model assumed that v can be expressed as a function of u, thus obtaining a closed system:

$$\partial_t u + \partial_x f(u) = 0, \tag{B.2}$$

where $u = u(t, x) \in [0, u_{max}]$, with u_{max} being the maximum density of cars on the road.

The usual assumptions on the functions $v = v(u)$ and $f = f(u) = u\, v(u)$ are the following:

1. v is decreasing.
2. f is monotonically increasing on an interval $[0, \sigma]$, $\sigma \in]0, u_{max}[$, and decreasing on $[\sigma, u_{max}]$.
3. σ is the unique maximum point of f.
4. f is concave.

On the interval $[0, \sigma]$, the traffic is said to be in free flow, while above σ the traffic is said to be in congested flow.

We define the following:

Definition 48 Let $\tau : [0, u_{max}] \rightarrow [0, u_{max}]$ be the map such that $f(\tau(u)) = f(u)$ for every $u \in [0, u_{max}]$ and $\tau(u) \neq u$ for every $u \in [0, u_{max}] \setminus \{\sigma\}$.

We are now ready to describe a model for LWR on a network. A network is a topological graph given by a couple $(\mathcal{I}, \mathcal{J})$, where $\mathcal{I} = \{I_i : i = 1, \cdots, N\}$ is a finite set of intervals parameterizing roads, and $\mathcal{J}$ is a finite set vertices representing junctions. We assume that on each I_i the evolution of car density is given by an LWR model (B.2), and thus the dynamics is specified if at junctions the evolution is uniquely determined.

It is easy to check that the conservation of cars through the junctions is not sufficient to isolate a unique solution, as first observed by Holden and Risebro [178]. The authors used the maximization of a functional at vertexes to get uniqueness of solutions, and then many alternative ideas were proposed (see [142] for a complete account). In order to implement such ideas, we first define the Riemann problem (RP) at a junction J, which is a Cauchy problem with initial data constant on each road incident at the junction. One can see that the evolution of traffic load on the whole network is assigned once one prescribes a Riemann solver at each junction, i.e., a map assigning a solution with every Riemann problem at the junction. Given initial conditions $(u_{i,0}, u_{j,0})$, where i runs on incoming roads and j on outgoing ones, density values $(\widehat{u}_i, \widehat{u}_j)$ are assigned so that the solution on the incoming road i is given by a single wave $(u_{i,0}, \widehat{u}_i)$ and on the outgoing road j by the single wave $(\widehat{u}_j, u_{j,0})$.

Let us first detail the concept of solution on a network. First, we assume that on each road (B.2) is satisfied in weak sense, that is, for every test function $\varphi = \varphi(t, x) : \mathbb{R}^+ \times I_i \to \mathbb{R}^+$, it holds

$$\int_0^{+\infty} \int_{a_i}^{b_i} \left(u_i \frac{\partial \varphi}{\partial t} + f(u_i) \frac{\partial \varphi}{\partial x} \right) \, dx \, dt = 0,$$

where $I_i = [a_i, b_i]$ and u_i is the car density on I_i. One further considers entropy conditions.

A solution at a junction J is defined as follows. Assume there are n incoming roads, say $I_1, \ldots, I_n$, and m outgoing ones, say $I_{n+1}, \ldots, I_{n+m}$, and then a solution is a collection of functions $u_1, \cdots, u_{n+m}$ such that

$$\sum_{l=1}^{n+m} \left[\int_0^{+\infty} \int_{a_l}^{b_l} \left(u_l \frac{\partial \varphi_l}{\partial t} + f(u_l) \frac{\partial \varphi_l}{\partial x} \right) \, dx \, dt \right] = 0 \qquad (B.3)$$

for every set of test functions φ_l smoothly connected at the junction, i.e., such that

$$\varphi_i(\cdot, b_i) = \varphi_j(\cdot, a_j), \qquad \frac{\partial \varphi_i}{\partial x}(\cdot, b_i) = \frac{\partial \varphi_j}{\partial x}(\cdot, a_j)$$

for all $i = 1, \ldots, n$ and all $j = n+1, \ldots, n+m$. A consequence of such definition is the following equality:

$$\sum_{i=1}^{n} f(u_i(t, b_i^-)) = \sum_{j=n+1}^{n+m} f(u_j(t, a_j^+)),$$

which implies the conservation of cars through the junction. We are now ready to detail the approaches to isolate a unique solution at the junction.

Recall that a *Riemann problem* for a conservation law on a real line consists of a Cauchy problem with Heaviside-type initial data, usually assumed to be constant to the left and right of the origin. See Sect. A.6 for a complete description. Self-similar solutions consist of shocks and rarefactions and solutions to general Cauchy problems are constructed via wave-front tracking using Riemann problems as building blocks, see Sect. A.7.1. The map associating a solution to every Heaviside-type initial datum is referred to as *Riemann solver*. Following the same logic, a Riemann problem at a junction is a Cauchy problem with constant initial data on every road, the junction representing the discontinuity point. Then, a Riemann solver at a junction J is defined as follows:

Definition 49 A *Riemann solver* (*RS*) at J is a mapping associating with every initial datum $u_0 = (u_{0,1}, \ldots, u_{0,n+m}) \in \mathbb{R}^{n+m}$ a vector $\hat{u} = (\hat{u}_1, \ldots, \hat{u}_{n+m}) \in \mathbb{R}^{n+m}$ such that:

(i) On every incoming road I_i, the solution is that of the Riemann problem $(u_{0,i}, \hat{u}_i)$.

(ii) On every outgoing road I_j, the solution is that of the Riemann problem $(\hat{u}_j, u_{0,j})$.

The Riemann solver must satisfy the consistency condition $RS(RS(u_0)) = RS(u_0)$.

To have a well-defined solution and conserve the number of cars through the junction, we further impose the following:

(H1)   The waves generated from the junction must have negative speeds on the incoming arcs and positive speeds on the outgoing ones.

(H2)   The solution to a Riemann problem at a vertex must satisfy Eq. (B.3).

(H3)   The mapping $u_{0,l} \mapsto f(\hat{u}_l)$ is continuous for every $l = 1, \cdots, n + m$.

(H1) ensured conservation of cars, while (H2) is necessary to define a weak solution at the junction.

If we assign a Riemann solver at every junction J, then a solution on the whole network will be a solution u to the conservation law on every road and such that

$$RS(u_J(t)) = u_J(t),$$

where $u_J(t) := (u_1(t, b_1^-), \ldots, u_n(t, b_n^-), u_{n+1}(t, a_{n+1}^+), \ldots, u_{n+m}(t, a_{n+m}^+)) \in \mathbb{R}^{n+m}$.

B.2   Dynamics at Simple Junctions

Various approaches were proposed in the literature to define Riemann Solvers at junctions. Most of them are based on the following rules:

(A) Traffic distribution coefficients $\alpha_{ji} \in]0, 1[$ represent the percentage of traffic moving from incoming road i to outgoing road j. Such coefficients can be organized in a traffic distribution matrix:

$$A = \left\{\alpha_{ji}\right\}_{j=n+1,\ldots,n+m, \ i=1,\ldots,n} \in \mathbb{R}^{m \times n}.$$

A is row stochastic, i.e., for every $i = 1, \cdots, n$,

$$\sum_{j=n+1}^{n+m} \alpha_{ji} = 1.$$

(B) Drivers behave so as to maximize the flux through the junction, while distributing according to rule (A).

If $n > m$, that is, if there are more incoming than outgoing roads, then an additional rule is needed:

(C) There is a priority vector $(p_1, \ldots, p_n)$ assigning the percentage of traffic through the junction from each road. For instance, if there are two incoming roads a and b and one outgoing road c and Q is the amount of cars going through the junction, then $p_1 Q$ comes from a and $p_2 Q = (1 - p_1) Q$ from b.

Following rules **(A)**, **(B)**, and **(C)**, we can uniquely define a Riemann solver as follows. First, we will define the solution by assigning an Initial-Boundary Value Problem (IBVP) on each road. These IBVPs will be designed so that the boundary value would be attained by the solution; otherwise, we may violate conservation of cars through the junction. This is achieved by imposing admissible values on each road generating waves with negative speed on incoming roads and positive on outgoing ones, see also [142] for an extensive discussion. Such restriction has the advantage of allowing to define only the flow through the junction, being the values of the densities automatically determined by the admissibility condition on waves' speed sign. We explain this fact below in detail for the case of simple junctions, but first we start proving it for the general case.

The achievable flow on each road is described by the following:

Proposition 13 *Let* $(u_{1,0}, u_{2,0}, \ldots, u_{n+m,0})$ *be the initial densities of an RP at J and* γ_i^{max}, $i = 1, \ldots, n$, *and* γ_j^{max}, $j = n+1, \ldots, n+m$, *be the maximum fluxes that can be obtained on incoming roads and outgoing roads, respectively. Then,*

$$
\gamma_i^{max} = \begin{cases} f\left(u_{i,0}\right), & \text{if } u_{i,0} \in [0, \sigma], \\ f\left(\sigma\right), & \text{if } u_{i,0} \in]\sigma, u_{max}], \end{cases} \quad i = 1, \cdots, n, \tag{B.4}
$$

$$
\gamma_j^{max} = \begin{cases} f\left(\sigma\right), & \text{if } u_{j,0} \in [0, \sigma], \\ f\left(u_{j,0}\right), & \text{if } u_{j,0} \in]\sigma, u_{max}], \end{cases} \quad j = n+1, \cdots, n+m. \tag{B.5}
$$

In particular, densities can be reconstructed by flows at the junction.

Proof We focus on incoming road, being the outgoing case similar. Fix an incoming road i, and let $\widehat{u}_i$ be the trace at the junction given by the Riemann solver and $(u_{i,0}, \widehat{u}_i)$ the corresponding wave with negative speed. If $u_{i,0} \in [0, \sigma]$, then $\widehat{u}_i \in \{u_{i,0}\} \cup]\tau(u_{i,0}), 1]$. Thus either there is no wave, if $\widehat{u}_i = u_{i,0}$, or the wave is a shock with negative speed, see Fig. B.1 (left), which would lower the flux. Therefore the maximal flux is given by $f(u_{i,0})$. Notice also that for every flux value in $[0, f(u_{i,0})]$, there exists a unique $\widehat{u}_i$.

If, instead, $u_{i,0} \in [\sigma, 1]$, then $\widehat{u}_i \in [\sigma, 1]$. The generated wave $(u_{i,0}, \widehat{u}_i)$ is a rarefaction or a shock with negative speed, see Fig. B.1 (right). Then every flux can be achieved, i.e., all of the interval $[0, f(\sigma)]$ and, again, there exists a unique admissible value of $\widehat{u}_i$ for each flux value.

The analysis for outgoing roads is similar, see Fig. B.2. □

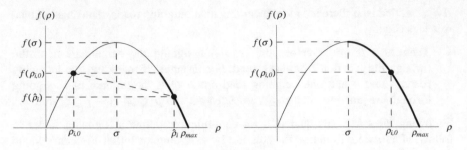

Fig. B.1 Images of Riemann solvers for the incoming roads

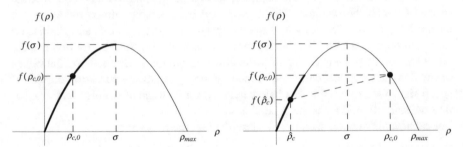

Fig. B.2 Images of Riemann Solvers for the outgoing road

Proposition 13 allows to restate rules **(A)**, **(B)**, and **(C)** as a linear programming problem in terms of the incoming fluxes $\widehat{\gamma}_i = f(\widehat{u}_i)$. Indeed, rule **(A)** allows to determine the outgoing fluxes $\widehat{\gamma}_j = f(\widehat{u}_j)$ in terms of the incoming ones. Then rule **(B)** provides a linear functional in the fluxes $\widehat{\gamma}_i$ to be maximized. The constraints are given by the formulas (B.4) and (B.5). Rule **(C)** allows to choose a unique solution to the linear programming problem in case of more incoming than outgoing roads.

In the following sections, we will explicitly solve the Riemann problems in the following cases: junctions of type 2×1 (two incoming roads and one outgoing road), junctions of type 1×2 (one incoming road and two outgoing roads), and junctions of type 2×2 (two incoming roads and two outgoing roads). We refer the reader to [142] for a complete description of the general case.

B.2.1   Two Incoming and One Outgoing Roads

Assume to have two incoming roads a and b and one outgoing road c. For constant initial data $(u_{a,0}, u_{b,0}, u_{c,0})$, the solution is defined as follows. Define

$$\widehat{\gamma}_c = \min\left\{\gamma_a^{max} + \gamma_b^{max}, \gamma_c^{max}\right\},$$

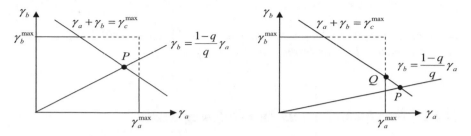

Fig. B.3 The possible different cases described in Sect. B.2.1

where γ_i^{max}, $i = a, b$, is defined as in (B.4) and γ_c^{max} as in (B.5). The quantity $\widehat{\gamma}_c$ is the maximal flux through the junction, thus respecting rule **(B)**. For rule **(A)**, the traffic distribution matrix is simply the vector $(1, 1)$ and no restriction is imposed. In the space of incoming fluxes (γ_a, γ_b) define a line by

$$\gamma_b = \frac{1-q}{q}\gamma_a. \tag{B.6}$$

Such a line reflects rule **(C)**, and we set P to be the point of intersection of the line (B.6) with the maximal flux line $\gamma_a + \gamma_b = \widehat{\gamma}_c$. If the point P belongs to the admissible region $\Omega = \{(\gamma_a, \gamma_b): 0 \leq \gamma_i \leq \gamma_i^{max}, \ 0 \leq \gamma_a + \gamma_b \leq \widehat{\gamma}_c, \ i = a, b\}$, then P is the solution; see Fig. B.3, left. Otherwise, we set $(\widehat{\gamma}_a, \widehat{\gamma}_b) = Q$, where Q is the point of $\Omega \cap \{(\gamma_a, \gamma_b) : \ \gamma_a + \gamma_b = \widehat{\gamma}_c\}$ closest to the line (B.6), see Fig. B.3, right. Given the fluxes, the densities are determined uniquely. The construction is summarized in the following:

Proposition 14 *Consider a junction J with $n = 2$ incoming roads and $m = 1$ outgoing road. For every $u_{a,0}, u_{b,0}, u_{c,0} \in [0, u_{max}]$, there exists a unique admissible weak solution $u = (u_a, u_b, u_c)$ at the junction J, satisfying rules **(A)**, **(B)**, and **(C)**, such that*

$$u_a\,(0, \cdot) \equiv u_{a,0}, \qquad u_b\,(0, \cdot) \equiv u_{b,0}, \qquad u_c\,(0, \cdot) \equiv u_{c,0}.$$

Moreover, there exists a unique 3-tuple $(\widehat{u}_a, \widehat{u}_b, \widehat{u}_c) \in [0, u_{max}]^3$ such that

$$\widehat{u}_i \in \begin{cases} \{u_{i,0}\} \cup \,]\tau\,(u_{i,0})\,, u_{max}], & if\,0 \leq u_{i,0} \leq \sigma, \\ [\sigma, u_{max}]\,, & if\,\sigma \leq u_{i,0} \leq u_{max}, \end{cases} \quad i = a, b,$$

and

$$\widehat{u}_c \in \begin{cases} [0, \sigma]\,, & if\,0 \leq u_{c,0} \leq \sigma, \\ \{u_{c,0}\} \cup [0, \tau\,(u_{c,0})[\,, & if\,\sigma \leq u_{c,0} \leq u_{max}, \end{cases}$$

and for $i \in \{a, b\}$, the solution is given by the wave $(u_{i,0}, \widehat{u}_i)$, while for the outgoing road, the solution is given by the wave $(\widehat{u}_c, u_{c,0})$.

B.2.2   One Incoming and Two Outgoing Roads

Let us now consider the junction with the incoming road a and two outgoing roads b and c. The distribution matrix A, of rule (**A**), takes the form

$$A = \begin{pmatrix} \alpha \\ 1 - \alpha \end{pmatrix},$$

where $\alpha \in \,]0, 1[$ and $(1 - \alpha)$ indicate the percentage of cars which, from road a, goes to roads b and c, respectively. Thanks to rule (**B**), the solution to an RP is

$$\widehat{\gamma} = (\widehat{\gamma}_a, \widehat{\gamma}_b, \widehat{\gamma}_c) = (\widehat{\gamma}_a, \alpha\widehat{\gamma}_a, (1 - \alpha)\,\widehat{\gamma}_a),$$

where

$$\widehat{\gamma}_a = \min\left\{\gamma_a^{max}, \frac{\gamma_b^{max}}{\alpha}, \frac{\gamma_c^{max}}{1 - \alpha}\right\}.$$

Once we have obtained $\widehat{\gamma}_a$, $\widehat{\gamma}_b$, and $\widehat{\gamma}_c$, it is possible to find in a unique way $\widehat{u}_i$, $i \in \{a, b, c\}$, reasoning as in the proof of Proposition 14. Then we obtain the following:

Proposition 15 *Consider a junction J with $n = 1$ incoming road and $m = 2$ outgoing roads. For every $u_{a,0}, u_{b,0}, u_{c,0} \in [0, u_{max}]$, there exists a unique admissible weak solution $u = (u_{a,b}, u_c)$ at the junction J, respecting rules (**A**) and (**B**), such that*

$$u_a(0, \cdot) \equiv u_{a,0}, \qquad u_b(0, \cdot) \equiv u_{b,0}, \qquad u_c(0, \cdot) \equiv u_{c,0}.$$

Moreover, there exists a unique 3-tuple $(\widehat{u}_a, \widehat{u}_b, \widehat{u}_c) \in [0, u_{max}]^3$ such that

$$\widehat{u}_a \in \begin{cases} \{u_{a,0}\} \cup \,]\tau(u_{a,0}), u_{max}], & \text{if } 0 \leq u_{a,0} \leq \sigma, \\ [\sigma, u_{max}], & \text{if } \sigma \leq u_{a,0} \leq u_{max}, \end{cases}$$

and

$$\widehat{u}_j \in \begin{cases} [0, \sigma], & \text{if } 0 \leq u_{j,0} \leq \sigma, \\ \{u_{j,0}\} \cup [0, \tau(u_{j,0})[, & \text{if } \sigma \leq u_{j,0} \leq u_{max}, \end{cases} \quad j = b, c,$$

and for the incoming road, the solution is given by the wave $(u_{a,0}, \widehat{u}_a)$, while for $j = b, c$, the solution is given by the wave $(\widehat{u}_j, u_{j,0})$.

B.2.3   Two Incoming and Two Outgoing Roads

Let us now consider the junction with two incoming roads a and b and two outgoing roads c and d. The distribution matrix A, of rule **(A)**, takes the form

$$A = \begin{pmatrix} \alpha & \beta \\ 1 - \alpha & 1 - \beta \end{pmatrix},$$ (B.7)

where $\alpha, \beta \in\,]0, 1[$. We assume that $\alpha \neq \beta$; otherwise, we may have more than one solution to the linear programming problem, see [142] for details.

First notice that constraints from outgoing roads fluxes can be expressed as

$$\alpha \widehat{\gamma}_a + \beta \widehat{\gamma}_b \le \gamma_c^{max}, \quad (1 - \alpha)\widehat{\gamma}_a + (1 - \beta)\widehat{\gamma}_b \le \gamma_d^{max}.$$

Define $P = (\gamma_1, \gamma_2)$ to be the point of intersection of the two lines:

$$\alpha \gamma_1 + \beta \gamma_2 = \gamma_c^{max}, \quad (1 - \alpha)\gamma_1 + (1 - \beta)\gamma_2 = \gamma_d^{max}.$$

To express the solution, we need to distinguish some cases:

Case a).   If $\gamma_1 \le \gamma_a^{max}$ and $\gamma_2 \le \gamma_b^{max}$, then the solution is given by

$$\widehat{\gamma}_a = \gamma_1, \quad \widehat{\gamma}_b = \gamma_2.$$

Case b).   If $\gamma_1 > \gamma_a^{max}$ and $\gamma_2 > \gamma_b^{max}$, then the solution is given by

$$\widehat{\gamma}_a = \gamma_a^{max}, \quad \widehat{\gamma}_b = \gamma_b^{max}.$$

Case c).   Assume $\gamma_1 > \gamma_a^{max}$ and $\gamma_2 \le \gamma_b^{max}$. If $\alpha < \beta$ (thus $1 - \beta < 1 - \alpha$), then the constraint given by outgoing road c is more stringent than that of outgoing road d, and thus the solution is given by

$$\widehat{\gamma}_a = \gamma_a^{max}, \quad \widehat{\gamma}_b = \min\left\{ \frac{\gamma_c^{max} - \alpha \gamma_a^{max}}{\beta}, \gamma_b^{max} \right\}.$$

Otherwise, i.e., if $\alpha > \beta$, then the solution is given by

$$\widehat{\gamma}_a = \gamma_a^{max}, \quad \widehat{\gamma}_b = \min\left\{ \frac{\gamma_d^{max} - (1 - \alpha)\gamma_a^{max}}{1 - \beta}, \gamma_b^{max} \right\}.$$

Case d).   Assume $\gamma_1 \leq \gamma_a^{max}$ and $\gamma_2 > \gamma_b^{max}$. If $\alpha > \beta$ (thus $1 - \beta > 1 - \alpha$), then the constraint given by outgoing road c is more stringent than that of outgoing road d, and thus the solution is given by

$$\widehat{\gamma_a} = \min\left\{\frac{\gamma_c^{max} - \beta\gamma_b^{max}}{\alpha}, \gamma_a^{max}\right\}, \quad \widehat{\gamma_b} = \gamma_b^{max}.$$

Otherwise, i.e., if $\alpha < \beta$, then the solution is given by

$$\widehat{\gamma_a} = \min\left\{\frac{\gamma_d^{max} - (1 - \beta)\gamma_b^{max}}{1 - \alpha}, \gamma_a^{max}\right\}, \quad \widehat{\gamma_b} = \gamma_b^{max}.$$

B.3   Constructing Solutions on a Network

In this section, we describe a general approach to construct solutions on networks. The scalar case was extensively studied and there are general results, while results for the systems case are available only in special cases.

Cauchy problems on a network are defined once initial data on each road are specified: $\mathbf{u}_0 = (u_{1,0}, \cdots, u_{N,0})$, where $u_{i,0} : I_i \rightarrow \mathbb{R}^d$ is a measurable and bounded function. In some cases, entering edges to the network are given, and then boundary conditions must be assigned.

A solution to the Cauchy problem on a network is an N vector-valued functions $u_i = u_i(t, x) : \mathbb{R}_+ \times I_i \rightarrow \mathbb{R}^d$ such that:

 (i) $t \mapsto \|u_i(t, \cdot)\|_{\mathbf{L}^1_{\text{loc}}}$ is continuous for each $i = 1, \cdots, N$.
 (ii) u_i is a weak entropic solution to (B.2) on I_i.
 (iii) At each junction, incoming and outgoing densities u_i give an admissible solution as specified in Sects. B.1 and B.2.
 (iv) $u_i(0, x) = u_{i,0}(x)$ for almost every $x \in I_i$.

Solutions can be constructed via wave-front tracking algorithm extended to the network case and consist of the following steps:

1. Approximate initial data with piecewise constant functions. Then, use wave-front tracking algorithm solving classical Riemann problems within roads and Riemann problems at junctions. At each interaction time, solve a new Riemann problem.
2. Estimate the number of waves to allow the construction for any time and the total variation of solution for compactness.
3. Using compactness properties, pass to the limit in approximations and prove the limit is a solution.

Notice that the number of waves may increase due to interactions with junctions, and thus suitable functionals must be used (see [139, 141, 142]). On the other side,

due to finite speed of propagation of waves, the total variation estimates on the flux
can be done separately for each junction.

Theorem 43 *For each junction J of the network, consider the network with only J
and incident roads prolonged to infinity. If there exists $C > 0$ such that for every
$\mathbf{u}_0 = (u_{1,0}, \cdots, u_{N,0})$ with bounded total variation we have*

$$\text{TV}\,(f(\mathbf{u}(t))) \leq C \cdot \text{TV}\,(f(\mathbf{u}_0)),$$

*then the total variation is bounded on the entire network (with a time-dependent
constant $C_t > 0$.)*

Once the estimate is available for the fluxes, still one has to deduce the same
estimate for the densities $\mathbf{u}_\nu$. General results are available for the scalar case and are
based on controlling the number of waves crossing the maximum of the flux called
big shocks. The general strategy works as follows. Define the following:

Definition 50 A wave (u_i^-, u_i^+) on road I_i is a *big shock* if $u_i^- < u_i^+$ and

$$\text{sgn}\,(u_i^- - \sigma) \cdot \text{sgn}\,(u_i^+ - \sigma) < 0.$$

Definition 51 Given a junction J and an incoming arc I_i, $u_i(t, b_i-)$ is called *good
datum* at time $t > 0$ if $u_i(t, b_i-) \in [\sigma, u_{\max}]$ and *bad datum* otherwise. For an
outgoing arc I_j, $u_j(t, a_j+)$ is a good datum at time $t > 0$ if

$$u_j(t, a_j+) \in [0, \sigma]$$

and a bad datum otherwise.

Then, we have the following lemma:

Lemma 15 *If an arc I_i incident to a junction J has a good datum, then the datum
remains good and no big shocks emerge from J until a wave interacts with J from
I_i. If I_i has a bad datum, then at an interaction time either no wave is produced on
I_i or a big shock is produced on I_i and the new density is a good datum.*

Lemma 15 allows to limit the number of big shocks on every road, and thus the flux
$u \mapsto f(u)$ can be inverted and total variation estimates on u hold true. We refer the
reader to [139, 142] for details. We also refer the reader to Sect. 3.2 for details on
Riemann solvers at junctions which guarantee existence of solutions.

Bibliography

1. Adimurthi, S. Ghoshal, and G. Veerappa Gowda. Exact controllability of scalar conservation laws with strict convex flux. *Mathematical Control and Related Fields*, 4(4):401–449, 2014. cited By 15.
2. S. Agarwal, P. Kachroo, S. Contreras, and S. Sastry. Feedback-coordinated ramp control of consecutive on-ramps using distributed modeling and Godunov-based satisfiable allocation. *IEEE Transactions on Intelligent Transportation Systems*, 16(5):2384–2392, 2015. cited By 12.
3. A. Alessandri, A. Di Febbraro, A. Ferrara, and E. Punta. Nonlinear optimization for freeway control using variable-speed signaling. *IEEE Transactions on vehicular technology*, 48(6):2042–2052, 1999.
4. G. Allaire. *Numerical analysis and optimization*. Oxford university press, 2007.
5. D. Amadori Initial-boundary value problems for nonlinear systems of conservation laws. *NoDEA: Nonlinear Differential Equations and Applications*, 4(1):1–42, 1997.
6. D. Amadori and R. Colombo. Continuous dependence for 2 × 2 conservation laws with boundary. *Journal of Differential Equations*, 138(2):229–266, 1997.
7. D. Amadori and R. Colombo. Viscosity solutions and standard Riemann semigroup for conservation laws with boundary. *Rendiconti del Seminario Matematico della Universita di Padova*, 99:219–245, 1998.
8. L. Ambrosio, N. Fusco, and D. Pallara. *Functions of bounded variation and free discontinuity problems*, volume 254. Clarendon Press Oxford, 2000.
9. S. Amin, F. Hante, and A. Bayen. On stability of switched linear hyperbolic conservation laws with reflecting boundaries. *Hybrid System: Computation and Control, Lecture Notes in Computer Science*, 4981:602–605, 2008, https://doi.org/10.1007/978-3-540-78929-1_44.
10. F. Ancona, A. Cesaroni, G. M. Coclite, and M. Garavello. On the optimization of conservation law models at a junction with inflow and flow distribution controls. *SIAM J. Control Optim.*, 56(5):3370–3403, 2018.
11. F. Ancona and G. Coclite. On the attainable set for Temple class systems with boundary controls. *SIAM J. Control Optim.*, 43(6):2166–2190 (electronic), 2005.
12. F. Ancona and A. Marson. On the attainable set for scalar nonlinear conservation laws with boundary control. *SIAM J. Control Optim.*, 36(1):290–312 (electronic), 1998.
13. B. Andreianov, C. Donadello, S. S. Ghoshal, and U. Razafison. On the attainable set for a class of triangular systems of conservation laws. *J. Evol. Equ.*, 15(3):503–532, 2015.

© The Author(s) 2022
A. Bayen et al., *Control Problems for Conservation Laws with Traffic Applications*,
PNLDE Subseries in Control 99, https://doi.org/10.1007/978-3-030-93015-8

14. B. Andreianov, C. Donàdello, U. Razafison, and M. D. Rosini. Analysis and approximation of one-dimensional scalar conservation laws with general point constraints on the flux. *J. Math. Pures Appl. (9)*, 116:309–346, 2018.

15. B. Andreianov, C. Donadello, and M. D. Rosini. Crowd dynamics and conservation laws with nonlocal constraints and capacity drop. *Math. Models Methods Appl. Sci.*, 24(13):2685–2722, 2014.

16. B. Andreianov, C. Donadello, and M. D. Rosini. A second-order model for vehicular traffics with local point constraints on the flow. *Math. Models Methods Appl. Sci.*, 26(4):751–802, 2016.

17. B. Andreianov, P. Goatin, and N. Seguin. Finite volume schemes for locally constrained conservation laws. *Numer. Math.*, 115(4):609–645, 2010. With supplementary material available online.

18. B. Andreianov, F. Lagoutière, N. Seguin, and T. Takahashi. Well-posedness for a one-dimensional fluid-particle interaction model. *SIAM J. Math. Anal.*, 46(2):1030–1052, 2014.

19. B. Andreianov, S. Sundar Ghoshal, and K. Koumatos. Non-controllability of the viscous Burgers equation and a detour into the well-posedness of unbounded entropy solutions to scalar conservation laws. working paper or preprint, Mar. 2020.

20. D. Armbruster, P. Degond, and C. Ringhofer. A model for the dynamics of large queuing networks and supply chains. *SIAM J. Appl. Math.*, 66(3):896–920 (electronic), 2006.

21. J.-P. Aubin, A. Bayen, and P. Saint-Pierre. Dirichlet problems for some Hamilton–Jacobi equations with inequality constraints. *SIAM journal on control and optimization*, 47(5):2348–2380, 2008.

22. J.-P. Aubin, A. Bayen, and P. Saint-Pierre. *Viability theory: new directions*. Springer Science & Business Media, 2011.

23. A. Aw and M. Rascle. Resurrection of "second order" models of traffic flow. *SIAM J. Appl. Math.*, 60(3):916–938, 2000.

24. D. S. Bale, R. J. Leveque, S. Mitran, and J. A. Rossmanith. A wave propagation method for conservation laws and balance laws with spatially varying flux functions. *SIAM J. Sci. Comput.*, 24(3):955–978, 2002.

25. A. Balogh and M. Krstic. Infinite dimensional backstepping-style feedback transformations for a heat equation with an arbitrary level of instability. *European journal of control*, 8(2):165–175, 2002.

26. M. K. Banda, M. Herty, and A. Klar. Gas flow in pipeline networks. *Netw. Heterog. Media*, 1(1):41–56 (electronic), 2006.

27. M. Bardi and I. Capuzzo-Dolcetta. *Optimal control and viscosity solutions of Hamilton-Jacobi-Bellman equations*. Springer Science & Business Media, 2008.

28. M. Bardi and L. Evans. On Hopf's formulas for solutions of Hamilton-Jacobi equations. *Nonlinear Analysis: Theory, Methods & Applications*, 8(11):1373–1381, 1984.

29. C. Bardos, A.-Y. LeRoux, and J.-C. Nédélec. First order quasilinear equations with boundary conditions. *Communications in partial differential equations*, 4(9):1017–1034, 1979.

30. E. Barron and R. Jensen. Semicontinuous Viscosity Solutions For Hamilton–Jacobi Equations With Convex Hamiltonians. *Communications in Partial Differential Equations*, 15(12):293–309, Jan. 1990.

31. L. D. Baskar, B. De Schutter, and H. Hellendoorn. Model-based predictive traffic control for intelligent vehicles: Dynamic speed limits and dynamic lane allocation. In *Intelligent Vehicles Symposium, 2008 IEEE*, pages 174–179. IEEE, 2008.

32. G. Bastin and J.-M. Coron. *Stability and Boundary Stabilisation of 1-D Hyperbolic Systems*. Number 88 in Progress in Nonlinear Differential Equations and Their Applications. Springer International, 2016.

33. G. Bastin and J.-M. Coron. A quadratic Lyapunov function for hyperbolic density-velocity systems with nonuniform steady states. *Systems & Control Letters*, 104:66–71, 2017.

34. G. Bastin, J.-M. Coron, and B. d'Andréa Novel. On Lyapunov stability of linearised Saint-Venant equations for a sloping channel. *Netw. Heterog. Media*, 4(2):177–187, 2009.

35. G. Bastin, J.-M. Coron, A. Hayat, and P. Shang. Exponential boundary feedback stabilization of a shock steady state for the inviscid Burgers equation. *Math. Models Methods Appl. Sci.*, 29(2):271–316, 2019.

36. A. Bayen, C. Claudel, and P. Saint-Pierre. Viability-based computations of solutions to the Hamilton-Jacobi-Bellman equation. In *International Workshop on Hybrid Systems: Computation and Control*, pages 645–649. Springer, 2007.

37. M. Ben-Akiva, D. Cuneo, M. Hasan, M. Jha, and Q. Yang. Evaluation of freeway control using a microscopic simulation laboratory. *Transportation Research Part C: Emerging Technologies*, 11(1):29–50, 2003. cited By 48.

38. S. Bianchini and A. Bressan. Vanishing viscosity solutions of nonlinear hyperbolic systems. *Annals of Mathematics*, 161(1):223–342, 2005.

39. S. Bianchini and R. M. Colombo. On the stability of the standard Riemann semigroup. *Proceedings of the American Mathematical Society*, 130(7):1961–1973, 2002.

40. S. Blandin, X. Litrico, M. L. Delle Monache, B. Piccoli, and A. Bayen. Regularity and Lyapunov stabilization of weak entropy solutions to scalar conservation laws. *IEEE Trans. Automat. Control*, 62(4):1620–1635, 2017.

41. R. Borsche, R. M. Colombo, and M. Garavello. On the coupling of systems of hyperbolic conservation laws with ordinary differential equations. *Nonlinearity*, 23(11):2749–2770, 2010.

42. R. Borsche, R. M. Colombo, and M. Garavello. Mixed systems: ODEs - balance laws. *J. Differential Equations*, 252(3):2311–2338, 2012.

43. R. Borsche, R. M. Colombo, and M. Garavello. On the interactions between a solid body and a compressible inviscid fluid. *Interfaces Free Bound.*, 15(3):381–403, 2013.

44. R. Borsche, R. M. Colombo, M. Garavello, and A. Meurer. Differential equations modeling crowd interactions. *J. Nonlinear Sci.*, 25(4):827–859, 2015.

45. B. Boutin, C. Chalons, F. Lagoutière, and P. G. LeFloch. Convergent and conservative schemes for nonclassical solutions based on kinetic relations. I. *Interfaces Free Bound.*, 10(3):399–421, 2008.

46. D. M. Bošković, A. Balogh, and M. Krstić. Backstepping in infinite dimension for a class of parabolic distributed parameter systems. *Math. Control Signals Systems*, 16(1):44–75, 2003.

47. A. Brandi, A. Ferrara, S. Sacone, S. Siri, C.Vivas, and F. R. Rubio. Model predictive control with state estimation for freeway systems. *Proceedings of the 2017 American Control Conference (ACC 2017), Seattle, WA, USA*, 2017.

48. A. Bressan. Contractive metrics for nonlinear hyperbolic systems. *Indiana Univ. Math. J.*, 37(2):409–421, 1988.

49. A. Bressan. A contractive metric for systems of conservation laws with coinciding shock and rarefaction curves. *J. Differential Equations*, 106(2):332–366, 1993.

50. A. Bressan. The unique limit of the Glimm scheme. *Arch. Rational Mech. Anal.*, 130(3):205–230, 1995.

51. A. Bressan. *Hyperbolic systems of conservation laws*, volume 20 of *Oxford Lecture Series in Mathematics and its Applications*. Oxford University Press, Oxford, 2000. The one-dimensional Cauchy problem.

52. A. Bressan and G. M. Coclite. On the boundary control of systems of conservation laws. *SIAM J. Control Optim.*, 41(2):607–622, 2002.

53. A. Bressan and R. M. Colombo. The semigroup generated by 2×2 conservation laws. *Archive for rational mechanics and analysis*, 133(1):1–75, 1995.

54. A. Bressan, G. Crasta, and B. Piccoli. Well-posedness of the Cauchy problem for $n \times n$ systems of conservation laws. *Mem. Amer. Math. Soc.*, 146(694):viii+134, 2000.

55. A. Bressan and B. Piccoli. *Introduction to the mathematical theory of control*. American Institute of Mathematical Sciences (AIMS), Springfield, MO, 2007.

56. G. Bretti, E. Cristiani, C. Lattanzio, A. Maurizi, and B. Piccoli. Two algorithms for a fully coupled and consistently macroscopic PDE-ODE system modeling a moving bottleneck on a road. *Mathematics in Engineering*, 1(1):55–83, 2019.

57. G. Bretti, R. Natalini, and B. Piccoli. Fast algorithms for a traffic flow model on networks. *Discrete and Continuous Dynamical Systems - Series B*, 6(3):427–448, 2006.

58. G. Bretti and B. Piccoli. A tracking algorithm for car paths on road networks. *SIAM J. Appl. Dyn. Syst.*, 7(2):510–531, 2008.

59. J. Burns and S. Kang. A control problem for Burgers equation with bounded input/output. *Nonlinear Dynamics*, 2(4):235–262, 1991.

60. C. Byrnes, D. Gilliam, and V. Shubov. On the global dynamics of a controlled viscous Burgers equation. *Journal of Dynamical and Control Systems*, 4(4):457–519, 1998.

61. E. Canepa and C. Claudel. Exact solutions to traffic density estimation problems involving the Lighthill-Whitham-Richards traffic flow model using mixed integer programming. In *2012 15th International IEEE Conference on Intelligent Transportation Systems*, pages 832–839. IEEE, 2012.

62. S. Čanić and E. H. Kim. Mathematical analysis of the quasilinear effects in a hyperbolic model blood flow through compliant axi-symmetric vessels. *Math. Methods Appl. Sci.*, 26(14):1161–1186, 2003.

63. R. C. Carlson, D. Manolis, I. Papamichail, and M. Papageorgiou. Integrated ramp metering and mainstream traffic flow control on freeways using variable speed limits. *Procedia - Social and Behavioral Sciences*, 48(0):1578–1588, 2012. Transport Research Arena 2012.

64. A. Cascone, C. D'Apice, B. Piccoli, and L. Rarità. Optimization of traffic on road networks. *Math. Models Methods Appl. Sci.*, 17(10):1587–1617, 2007.

65. A. Cascone, R. Manzo, B. Piccoli, and L. Rarità. Optimization versus randomness for car traffic regulation. *Phys. Rev. E*, 78:026113, Aug 2008.

66. A. Cascone, B. Piccoli, and L. Rarita. Circulation of car traffic in congested urban areas. *Communications in Mathematical Sciences*, 6(3):765–784, 2008.

67. C. Chalons, M. L. Delle Monache, and P. Goatin. A conservative scheme for non-classical solutions to a strongly coupled PDE-ODE problem. *Interfaces Free Bound.*, 19(4):553–570, 2017.

68. C. Chalons, P. Goatin, and N. Seguin. General constrained conservation laws. Application to pedestrian flow modeling. *Netw. Heterog. Media*, 8(2):433–463, 2013.

69. M. Chapouly. Global controllability of nonviscous and viscous Burgers-type equations. *SIAM J. Control Optim.*, 48(3):1567–1599, 2009.

70. Y. Chitour and B. Piccoli. Traffic circles and timing of traffic lights for cars flow. *Discrete Contin. Dyn. Syst. Ser. B*, 5(3):599–630, 2005.

71. F. Clarke, Y. Ledyaev, R. Stern, and P. Wolenski. *Nonsmooth analysis and control theory*, volume 178. Springer Science & Business Media, 2008.

72. C. Claudel and A. Bayen. Lax–Hopf based incorporation of internal boundary conditions into Hamilton–Jacobi equation. part i: Theory. *IEEE Transactions on Automatic Control*, 55(5):1142–1157, 2010.

73. C. Claudel and A. Bayen. Lax–Hopf based incorporation of internal boundary conditions into Hamilton-Jacobi equation. part ii: Computational methods. *IEEE Transactions on Automatic Control*, 55(5):1158–1174, 2010.

74. C. Claudel and A. Bayen. Convex formulations of data assimilation problems for a class of Hamilton-Jacobi equations. *SIAM Journal on Control and Optimization*, 49(2):383–402, 2011.

75. G. M. Coclite and M. Garavello. Vanishing viscosity for mixed systems with moving boundaries. *J. Funct. Anal.*, 264(7):1664–1710, 2013.

76. G. M. Coclite, M. Garavello, and B. Piccoli. Traffic flow on a road network. *SIAM J. Math. Anal.*, 36(6):1862–1886 (electronic), 2005.

77. R. M. Colombo and M. Garavello. A well posed Riemann problem for the p-system at a junction. *Netw. Heterog. Media*, 1(3):495–511 (electronic), 2006.

78. R. M. Colombo and P. Goatin. A well posed conservation law with a variable unilateral constraint. *J. Differential Equations*, 234(2):654–675, 2007.

79. R. M. Colombo, P. Goatin, and M. D. Rosini. On the modelling and management of traffic. *ESAIM Math. Model. Numer. Anal.*, 45(5):853–872, 2011.

80. R. M. Colombo and A. Groli. Minimising stop and go waves to optimise traffic flow. *Appl. Math. Lett.*, 17(6):697–701, 2004.
81. R. M. Colombo and A. Marson. Conservation laws and O.D.E.s. A traffic problem. In *Hyperbolic problems: theory, numerics, applications*, pages 455–461. Springer, Berlin, 2003.
82. R. M. Colombo and A. Marson. A Hölder continuous ODE related to traffic flow. *Proc. Roy. Soc. Edinburgh Sect. A*, 133(4):759–772, 2003.
83. R. M. Colombo and E. Rossi. IBVPs for scalar conservation laws with time discontinuous fluxes. *Math. Methods Appl. Sci.*, 41(4):1463–1479, 2018.
84. J.-M. Coron. Global asymptotic stabilization for controllable systems without drift. *Math. Control Signals Systems*, 5(3):295–312, 1992.
85. J.-M. Coron. Some open problems in control theory. In *Differential geometry and control (Boulder, CO, 1997)*, volume 64 of *Proc. Sympos. Pure Math.*, pages 149–162. Amer. Math. Soc., Providence, RI, 1999.
86. J.-M. Coron. Some open problems on the control of nonlinear partial differential equations. In *Perspectives in nonlinear partial differential equations*, volume 446 of *Contemp. Math.*, pages 215–243. Amer. Math. Soc., Providence, RI, 2007.
87. J.-M. Coron and G. Bastin. Dissipative boundary conditions for one-dimensional quasi-linear hyperbolic systems: Lyapunov stability for the C^1-norm. *SIAM Journal on Control and Optimization*, 53(3):1464–1483, 2015.
88. J.-M. Coron, G. Bastin, and B. d'Andréa Novel. Dissipative boundary conditions for one-dimensional nonlinear hyperbolic systems. *SIAM J. Control Optim.*, 47(3):1460–1498, 2008.
89. J.-M. Coron and B. d'Andréa Novel. Stabilization of a rotating body beam without damping. *IEEE Trans. Automat. Control*, 43(5):608–618, 1998.
90. J.-M. Coron, B. d'Andrea-Novel, and G. Bastin. A strict Lyapunov function for boundary control of hyperbolic systems of conservation laws. *IEEE Transactions on Automatic Control*, 52(1):2–11, 2007.
91. J.-M. Coron, S. Ervedoza, S. S. Ghoshal, O. Glass, and V. Perrollaz. Dissipative boundary conditions for 2×2 hyperbolic systems of conservation laws for entropy solutions in BV. *J. Differential Equations*, 262(1):1–30, 2017.
92. J.-M. Coron and A. Hayat. PI controllers for 1-D nonlinear transport equation. working paper or preprint, 2018.
93. J.-M. Coron and H.-M. Nguyen. Dissipative boundary conditions for nonlinear 1-D hyperbolic systems: sharp conditions through an approach via time-delay systems. *SIAM J. Math. Anal.*, 47(3):2220–2240, 2015.
94. M. Crandall, L. Evans, and P.-L. Lions. Some properties of viscosity solutions of Hamilton-Jacobi equations. *Transactions of the American Mathematical Society*, 282(2):487–502, 1984.
95. M. Crandall and P.-L. Lions. Viscosity solutions of Hamilton-Jacobi equations. *Transactions of the American mathematical society*, 277(1):1–42, 1983.
96. A. Csikos, I. Varga, and K. M. Hangos. Freeway shockwave control using ramp metering and variable speed limits. *21st Mediterranean Conference on Control & Automation*, pages 1569–1574, 2013.
97. A. Cutolo, C. D'Apice, and R. Manzo. Traffic optimization at junctions to improve vehicular flows. *ISRN Appl. Math.*, pages Art. ID 679056, 19, 2011.
98. C. M. Dafermos. Generalized characteristics and the structure of solutions of hyperbolic conservation laws. *Indiana Univ. Math. J.*, 26(6):1097–1119, 1977.
99. C. M. Dafermos. *Hyperbolic conservation laws in continuum physics*, volume 325 of *Grundlehren der Mathematischen Wissenschaften [Fundamental Principles of Mathematical Sciences]*. Springer-Verlag, Berlin, second edition, 2005.
100. C. Daganzo. The cell transmission model: A dynamic representation of highway traffic consistent with the hydrodynamic theory. *Transportation Research Part B*, 28:269–287, 1994.
101. C. F. Daganzo and J. A. Laval. Moving bottlenecks: A numerical method that converges in flows. *Transportation Research Part B: Methodological*, 39(9):855–863, 2005.
102. C. F. Daganzo and J. A. Laval. On the numerical treatment of moving bottlenecks. *Transportation Research Part B: Methodological*, 39(1):31–46, 2005.

103. R. Dager and E. Zuazua. *Wave Propagation, Observation and Control in 1-d Flexible Multi-Structures*. Springer, Berlin, Heidelberg, 2006.

104. E. Dal Santo, C. Donadello, S. F. Pellegrino, and M. D. Rosini. Representation of capacity drop at a road merge via point constraints in a first order traffic model. *ESAIM Math. Model. Numer. Anal.*, 53(1):1–34, 2019.

105. E. Dal Santo, M. D. Rosini, N. Dymski, and M. Benyahia. General phase transition models for vehicular traffic with point constraints on the flow. *Math. Methods Appl. Sci.*, 40(18):6623–6641, 2017.

106. C. D'Apice, S. Göttlich, M. Herty, and B. Piccoli. *Modeling, simulation, and optimization of supply chains*. Society for Industrial and Applied Mathematics (SIAM), Philadelphia, PA, 2010. A continuous approach.

107. C. D'Apice and R. Manzo. A fluid dynamic model for supply chains. *Netw. Heterog. Media*, 1(3):379–398 (electronic), 2006.

108. C. D'Apice, R. Manzo, and B. Piccoli. Packet flow on telecommunication networks. *SIAM J. Math. Anal.*, 38(3):717–740 (electronic), 2006.

109. L. C. Davis. Effect of adaptive cruise control systems on traffic flow. *Phys. Rev. E*, 69:066110, Jun 2004.

110. J. de Halleux, C. Prieur, J.-M. Coron, B. d'Andréa Novel, and G. Bastin. Boundary feedback control in networks of open channels. *Automatica J. IFAC*, 39(8):1365–1376, 2003.

111. M. L. Delle Monache and P. Goatin. A front tracking method for a strongly coupled PDE-ODE system with moving density constraints in traffic flow. *Discrete Contin. Dyn. Syst. Ser. S*, 7(3):435–447, 2014.

112. M. L. Delle Monache and P. Goatin. Scalar conservation laws with moving constraints arising in traffic flow modeling: an existence result. *J. Differential Equations*, 257(11):4015–4029, 2014.

113. M. L. Delle Monache and P. Goatin. A numerical scheme for moving bottlenecks in traffic flow. *Bull. Braz. Math. Soc. (N.S.)*, 47(2):605–617, 2016. Joint work with C. Chalons.

114. M. L. Delle Monache and P. Goatin. Stability estimates for scalar conservation laws with moving flux constraints. *Netw. Heterog. Media*, 12(2):245–258, 2017.

115. M. L. Delle Monache, T. Liard, B. Piccoli, R. Stern, and D. Work. Traffic reconstruction using autonomous vehicles. *SIAM J. Appl. Math.*, 79(5):1748–1767, 2019.

116. M. L. Delle Monache, B. Piccoli, and F. Rossi. Traffic regulation via controlled speed limit. *SIAM Journal on Control and Optimization*, 55(5):2936–2958, 2017.

117. M. L. Delle Monache, J. Reilly, S. Samaranayake, W. Krichene, P. Goatin, and A. M. Bayen. A PDE-ODE model for a junction with ramp buffer. *SIAM Journal on Applied Mathematics*, 74(1):22–39, 2014.

118. F. Di Meglio, R. Vazquez, and M. Krstic. Stabilization of a system of $n+1$ coupled first-order hyperbolic linear PDEs with a single boundary input. *IEEE Trans. Automat. Control*, 58(12):3097–3111, 2013.

119. J. I. Diaz. Obstruction and some approximate controllability results for the Burgers equation and related problems. In *Control of partial differential equations and applications (Laredo, 1994)*, volume 174 of *Lecture Notes in Pure and Appl. Math.*, pages 63–76. Dekker, New York, 1996.

120. M. Dick, M. Gugat, and G. Leugering. Classical solutions and feedback stabilization for the gas flow in a sequence of pipes. *Networks & Heterogeneous Media*, 5(4):691, 2010.

121. J. R. Domíngeuz Frejo and E. F. Camacho. Global versus local MPC algorithms in freeway traffic control with ramp metering and variable speed limits. *IEEE Transactions on intelligent transportation systems*, 13(4):1556–1565, 2012.

122. V. Dos Santos, G. Bastin, J.-M. Coron, and B. d'Andréa-Novel. Boundary control with integral action for hyperbolic systems of conservation laws: Stability and experiments. *Automatica*, 44(5):1310–1318, 2008.

123. F. Dubois and P. LeFloch. Boundary conditions for nonlinear hyperbolic systems of conservation laws. *J. Differential Equations*, 71(1):93–122, 1988.

124. N. S. Dymski, P. Goatin, and M. D. Rosini. Existence of **BV** solutions for a non-conservative constrained Aw-Rascle-Zhang model for vehicular traffic. *J. Math. Anal. Appl.*, 467(1):45–66, 2018.

125. N.S. Dymski, P. Goatin and M.D. Rosini, Modeling moving bottlenecks on road networks. In *Hyperbolic Problems: Theory, Numerics, Applications, AIMS on Applied Mathematics, Proceedings of the XVII international conference in Penn State*, June 2018, 10:419–426, 2020.

126. L. Evans. *Partial differential equations*, volume 19 of *Graduate Studies in Mathematics*. American Mathematical Society, Providence, RI, 2007.

127. L. C. Evans and R. F. Gariepy. *Measure Theory and Fine Properties of Functions*. CRC, 1991.

128. S. Fan, M. Herty, and B. Seibold. Comparative model accuracy of a data-fitted generalized Aw-Rascle-Zhang model. *Netw. Heterog. Media*, 9(2):239–268, 2014.

129. M. A. Fernández, V. Milišić, and A. Quarteroni. Analysis of a geometrical multiscale blood flow model based on the coupling of ODEs and hyperbolic PDEs. *Multiscale Model. Simul.*, 4(1):215–236, 2005.

130. E. Fernández-Cara and S. Guerrero. Remarks on the null controllability of the Burgers equation. *C. R. Math. Acad. Sci. Paris*, 341(4):229–232, 2005.

131. A. Ferrara, S. Sacone, and S. Siri. Event-triggered model predictive schemes for freeway traffic control. *Transportation Research Part C: Emerging Technologies*, 58:554–567, 2015.

132. A. F. Filippov. Differential equations with multi-valued discontinuous right-hand side. *Dokl. Akad. Nauk SSSR*, 151:65–68, 1963.

133. L. Formaggia, A. Quarteroni, and A. Veneziani. The circulatory system: from case studies to mathematical modeling. In *Complex systems in biomedicine*, pages 243–287. Springer Italia, Milan, 2006.

134. H. Frankowska. Lower semicontinuous solutions of Hamilton–Jacobi–Bellman equations. *SIAM J. Control Optim.*, 31:257–272, 1993.

135. A. V. Fursikov and O. Y. Imanuvilov. On controllability of certain systems simulating a fluid flow. In *Flow control (Minneapolis, MN, 1992)*, volume 68 of *IMA Vol. Math. Appl.*, pages 149–184. Springer, New York, 1995.

136. A. V. Fursikov and O. Y. Imanuvilov. Local exact controllability of the Navier-Stokes equations. *C. R. Acad. Sci. Paris Sér. I Math.*, 323(3):275–280, 1996.

137. M. Garavello and P. Goatin. The Aw-Rascle traffic model with locally constrained flow. *J. Math. Anal. Appl.*, 378(2):634–648, 2011.

138. M. Garavello, P. Goatin, T. Liard, and B. Piccoli. A multiscale model for traffic regulation via autonomous vehicles. *Journal of Differential Equations*, 296:6088–6124, 2020.

139. M. Garavello, K. Han, and B. Piccoli. *Models for vehicular traffic on networks*, volume 9 of *AIMS Series on Applied Mathematics*. American Institute of Mathematical Sciences (AIMS), Springfield, MO, 2016.

140. M. Garavello, R. Natalini, B. Piccoli, and A. Terracina. Conservation laws with discontinuous flux. *Netw. Heterog. Media*, 2(1):159–179, 2007.

141. M. Garavello and B. Piccoli. Traffic flow on a road network using the Aw-Rascle model. *Comm. Partial Differential Equations*, 31(1-3):243–275, 2006.

142. M. Garavello and B. Piccoli. *Traffic flow on networks*, volume 1 of *AIMS Series on Applied Mathematics*. American Institute of Mathematical Sciences (AIMS), Springfield, MO, 2006. Conservation laws models.

143. M. Garavello and B. Piccoli. Conservation laws on complex networks. *Ann. Inst. H. Poincaré Anal. Non Linéaire*, 26(5):1925–1951, 2009.

144. M. Garavello and B. Piccoli. Time-varying Riemann solvers for conservation laws on networks. *J. Differential Equations*, 247(2):447–464, 2009.

145. M. Garavello and S. Villa. The Cauchy problem for the Aw-Rascle-Zhang traffic model with locally constrained flow. *J. Hyperbolic Differ. Equ.*, 14(3):393–414, 2017.

146. I. Gasser, C. Lattanzio, and A. Maurizi. Vehicular traffic flow dynamics on a bus route. *Multiscale Model. Simul.*, 11(3):925–942, 2013.

147. V. Gayah and C. Daganzo. Analytical capacity comparison of one-way and two-way signalized street networks. *Transportation Research Record*, 2301:76–85, 2012. cited By 33.

148. M. Giles and S. Ulbrich. Convergence of linearized and adjoint approximations for discontinuous solutions of conservation laws. Part 2: Adjoint approximations and extensions. *SIAM Journal on Numerical Analysis*, 48(3):905–921, 2010.

149. O. Glass. On the controllability of the 1-d isentropic euler equation. *J. Eur. Math. Soc.(JEMS)*, 9(3):427–486, 2007.

150. O. Glass. On the controllability of the non-isentropic 1-d euler equation. *Journal of Differential Equations*, 257(3):638–719, 2014.

151. O. Glass and S. Guerrero. On the uniform controllability of the Burgers equation. *SIAM J. Control Optim.*, 46(4):1211–1238, 2007.

152. J. Glimm. Solutions in the large for nonlinear hyperbolic systems of equations. *Communication on pure and applied mathematics*, 18:697–715, 1965.

153. P. Goatin, S. Göttlich, and O. Kolb. Speed limit and ramp meter control for traffic flow networks. *Engineering Optimization*, 48(7):1121–1144, 2016.

154. S. K. Godunov. A difference method for numerical calculation of discontinuous solutions of the equations of hydrodynamics. *Mat. Sb. (N.S.)*, 47 (89):271–306, 1959.

155. G. Gomes and R. Horowitz. Optimal freeway ramp metering using the asymmetric cell transmission model. *Transportation Research Part C: Emerging Technologies*, 14(4):244–262, 2006.

156. S. Göttlich, M. Herty, and A. Klar. Modelling and optimization of supply chains on complex networks. *Commun. Math. Sci.*, 4(2):315–330, 2006.

157. J. M. Greenberg. Extensions and amplifications of a traffic model of Aw and Rascle. *SIAM Journal on Applied Mathematics*, 62(3):729–745, 2001.

158. J. M. Greenberg and T. Li. The effect of boundary damping for the quasilinear wave equation. *J. Differential Equations*, 52(1):66–75, 1984.

159. B. Greenshields. A study of traffic capacity. *Proceedings of the Highway Research Board*, 14:448–477, 1935.

160. M. Gugat, M. Herty, A. Klar, and Leugering. Optimal control for traffic flow networks. *Journal of optimization theory and applications*, 126(3):589–616, 2005.

161. M. Gugat and G. Leugering. Global boundary controllability of the Saint-Venant system for sloped canals with friction. In *Annales de l'Institut Henri Poincare*, volume 26, pages 257–270. Elsevier, 2009.

162. M. Gugat, G. Leugering, S. Tamasoiu, and K. Wang. H^2-stabilization of the isothermal Euler equations: a Lyapunov function approach. *Chinese Annals of Mathematics, Series B*, 33(4):479–500, 2012.

163. M. Guériau, R. Billot, N.-E. E. Faouzi], J. Monteil, F. Armetta, and S. Hassas. How to assess the benefits of connected vehicles? A simulation framework for the design of cooperative traffic management strategies. *Transportation Research Part C: Emerging Technologies*, 67:266–279, 2016.

164. J. K. Hale, S. M. V. Lunel, L. S. Verduyn, and S. M. V. Lunel. *Introduction to functional differential equations*, volume 99. Springer Science & Business Media, 1993.

165. K. Han and V. Gayah. Continuum signalized junction model for dynamic traffic networks: Offset, spillback, and multiple signal phases. *Transportation Research Part B: Methodological*, 77:213–239, 2015. cited By 22.

166. K. Han, V. Gayah, B. Piccoli, T. Friesz, and T. Yao. On the continuum approximation of the on-and-off signal control on dynamic traffic networks. *Transportation Research Part B: Methodological*, 61:73–97, 2014. cited By 34.

167. Y. Han, D. Chen, and S. Ahn. Variable speed limit control at fixed freeway bottlenecks using connected vehicles. *Transportation Research Part B: Methodological*, 98:113–134, 2017.

168. Y. Han, D. Chen, and S. Ahn. Variable speed limit control at fixed freeway bottlenecks using connected vehicles. *Transportation Research Part B: Methodological*, 98:113–134, 2017.

169. B. Haut and G. Bastin. A second order model of road junctions in fluid models of traffic networks. *Networks and Heterogeneous Media*, 2(2):227–253, 2007.

170. A. Hayat. Boundary Stability of 1-D Nonlinear Inhomogeneous Hyperbolic Systems for the C^1 Norm. *SIAM Journal on Control and Optimization*, 57(6):3603–3638, 2019.

171. A. Hayat and P. Shang. A quadratic Lyapunov function for Saint-Venant equations with arbitrary friction and space-varying slope. *Automatica J. IFAC*, 100:52–60, 2019.

172. Z. He, L. Zheng, L. Song, and N. Zhu. A jam-absorption driving strategy for mitigating traffic oscillations. *IEEE Transactions on Intelligent Transportation Systems*, 18(4):802–813, 2017.

173. A. Hegyi, B. De Schutter, and H. Hellendoorn. Model predictive control for optimal coordination of ramp metering and variable speed limits. *Transportation Research Part C*, 13(3):185–209, 2005.

174. A. Hegyi, B. D. Schutter, and J. Hellendoorn. Optimal coordination of variable speed limits to suppress shock waves. *IEEE Transactions on Intelligent Transportation Systems*, 6(1):102–112, 2005.

175. M. Herty, C. Kirchner, and S. Moutari. Multi-class traffic models on road networks. *Commun. Math. Sci.*, 4(3):591–608, 2006.

176. M. Herty and A. Klar. Modeling, simulation, and optimization of traffic flow networks. *SIAM J. Sci. Comput.*, 25(3):1066–1087, 2003.

177. M. Herty, S. Moutari, and M. Rascle. Optimization criteria for modelling intersections of vehicular traffic flow. *Netw. Heterog. Media*, 1(2):275–294 (electronic), 2006.

178. H. Holden and N. H. Risebro. A mathematical model of traffic flow on a network of unidirectional roads. *SIAM J. Math. Anal.*, 26(4):999–1017, 1995.

179. T. Horsin. On the controllability of the Burgers equation. *ESAIM Control Optim. Calc. Var.*, 3:83–95, 1998.

180. L. Hu, F. Di Meglio, R. Vazquez, and M. Krstic. Control of homodirectional and general heterodirectional linear coupled hyperbolic PDEs. *IEEE Trans. Automat. Control*, 61(11):3301–3314, 2016.

181. G.-R. Iordanidou, C. Roncoli, I. Papamichail, and M. Papageorgiou. Feedback-based mainstream traffic flow control for multiple bottlenecks on motorways. *IEEE Transactions on Intelligent Transportation Systems*, 16(2):610–621, 2015.

182. Z. Jia, C. Chen, B. Coifman, and P. Varaiya. The PeMS algorithms for accurate, real-time estimates of g-factors and speeds from single-loop detectors. In *ITSC 2001. 2001 IEEE Intelligent Transportation Systems. Proceedings (Cat. No. 01TH8585)*, pages 536–541. IEEE, 2001.

183. W.-L. Jin, Q.-J. Gan, and J.-P. Lebacque. A kinematic wave theory of capacity drop. *Transportation Research Part B: Methodological*, 81, Part 1:316–329, 2015.

184. K. Joseph and G. Gowda. Explicit formula for the solution of convex conservation laws with boundary condition. *Duke Math. J.*, 62(2):401–416, 03 1991.

185. K. Joseph and G. Gowda. Solution of convex conservation laws in a strip. *Proceedings of the Indian Academy of Sciences - Mathematical Sciences*, 102(1):29–47, Apr 1992.

186. I. Karafyllis and M. Papageorgiou. Feedback control of scalar conservation laws with application to density control in freeways by means of variable speed limits. *Automatica*, 105:228–236, 2019.

187. K. H. Karlsen, N. H. Risebro, and J. D. Towers. L^1 stability for entropy solutions of nonlinear degenerate parabolic convection-diffusion equations with discontinuous coefficients. *Skr. K. Nor. Vidensk. Selsk.*, (3):1–49, 2003.

188. C. T. Kelley. *Iterative methods for optimization.* Society for InÉhematics, Philadelphia, 1999.

189. B. Kerner and P. Konhäuser. Cluster effect in initially homogeneous traffic flow. *Physical Review E*, 48(4):R2335–R2338, 1993. cited By 339.

190. B. S. Kerner. *The physics of traffic: empirical freeway pattern features, engineering applications, and theory.* Springer, 2012.

191. B. Khondaker and L. Kattan. Variable speed limit: an overview. *Transportation Letters*, 7(5):264–278, 2015.

192. T. Kobayashi. Adaptive regulator design of a viscous Burgers system by boundary control. *IMA Journal of Mathematical Control and Information*, 18(3):427, 2001.

193. O. Kolb, S. Göttlich, and P. Goatin. Capacity drop and traffic control for a second order traffic model. *Networks & Heterogeneous Media*, 12(4):663–681, 2017.

194. A. Kotsialos, M. Papageorgiou, C. Diakaki, Y. Pavlis, and F. Middelham. Traffic flow modeling of large-scale motorway networks using the macroscopic modeling tool METANET. *IEEE Transactions on Intelligent Transportation Systems*, 3(4):282–292, 2002.

195. M. Krstic. On global stabilization of Burgers equation by boundary control. *Systems and Control Letters*, 37(3):123–141, 1999.

196. M. Krstic and A. Smyshlyaev. Backstepping boundary control for first-order hyperbolic PDEs and application to systems with actuator and sensor delays. *Systems Control Lett.*, 57(9):750–758, 2008.

197. M. Krstic and A. Smyshlyaev. *Boundary Control of PDEs: A Course on Backstepping Designs*, volume 16 of *Advances in Design and Control*. Society for Industrial and Applied Mathematics (SIAM), Philadelphia, PA, 2008.

198. S. Kruzhkov. First order quasilinear equations in several independent variables. *Mathematics of the USSR-Sbornik*, 10(2):217–243, 1970.

199. L. D. Landau and E. M. Lifschitz. *Lehrbuch der theoretischen Physik ("Landau-Lifschitz")*. Band VI. Akademie-Verlag, Berlin, fifth edition, 1991. Hydrodynamik. [Hydrodynamics], Translated from the Russian by Wolfgang Weller and Adolf Kühnel, Translation edited by Weller and with a foreword by Weller and P. Ziesche.

200. C. Lattanzio, A. Maurizi, and B. Piccoli. Moving bottlenecks in car traffic flow: a PDE-ODE coupled model. *SIAM J. Math. Anal.*, 43(1):50–67, 2011.

201. N. Laurent-Brouty, G. Costeseque, and P. Goatin. A coupled PDE-ODE model for bounded acceleration in macroscopic traffic flow models. *IFAC-PapersOnLine*, 51(9):37–42, 2018.

202. N. Laurent-Brouty, G. Costeseque, and P. Goatin. A macroscopic traffic flow model accounting for bounded acceleration, *SIAM J. Appl. Math.*, 81(1):173–189, 2021.

203. J. Laval. Stochastic processes of moving bottlenecks: Approximate formulas for highway capacity. *Transportation Research Record: Journal of the Transportation Research Board*, 1988:86–91, 2006.

204. M. Leautaud. Uniform controllability of scalar conservation laws in the vanishing viscosity limit. *SIAM Journal on Control and Optimization*, 50(3):1661–1699, 2012.

205. J. Lebacque and M. Khoshyaran. First order macroscopic traffic flow models for networks in the context of dynamic assignment. In M. Patriksson and M. Labbé, editors, *Transportation Planning*, volume 64 of *Applied Optimization*, pages 119–140. Springer US, 2002.

206. J.-P. Lebacque. The Godunov scheme and what it means for first order macroscopic traffic flow models. In *Proceedings of the 13th International Symposium on Transportation and Traffic Theory*, pages 647–677, Lyon, France, 1996.

207. J. P. Lebacque and M. M. Khoshyaran. Modelling vehicular traffic flow on networks using macroscopic models. In *Finite volumes for complex applications II*, pages 551–558. Hermes Sci. Publ., Paris, 1999.

208. J.-P. Lebacque, J. B. Lesort, and F. Giorgi. Introducing buses into first-order macroscopic traffic flow models. *Transportation Research Record*, 1644:70–79, 1998.

209. L. Leclercq. Bounded acceleration close to fixed and moving bottlenecks. *Transportation Research Part B: Methodological*, 41(3):309 – 319, 2007.

210. L. Leclercq, S. Chanut, and J.-B. Lesort. Moving bottlenecks in Lighthill-Whitham-Richards model: A unified theory. *Transportation Research Record: Journal of the Transportation Research Board*, 1883:3–13, 2004.

211. L. Leclercq, V. L. Knoop, F. Marczak, and S. P. Hoogendoorn. Capacity drops at merges: New analytical investigations. *Transportation Research Part C: Emerging Technologies*, 62:171 – 181, 2016.

212. P. LeFloch. Explicit formula for scalar nonlinear conservation laws with boundary condition. *Math. Methods Appl. Sci.*, 10(3):265–287, 1988.

213. P. G. LeFloch. *Hyperbolic systems of conservation laws*. Lectures in Mathematics ETH Zürich. Birkhäuser Verlag, Basel, 2002. The theory of classical and nonclassical shock waves.

214. H. Lenz, R. Sollacher, and M. Lang. Nonlinear speed-control for a continuum theory of traffic flow. *IFAC Proceedings Volumes*, 32(2):8339–8344, 1999.

215. G. Leugering and E. J. P. G. Schmidt. On the modelling and stabilization of flows in networks of open canals. *SIAM J. Control Optim.*, 41(1):164–180, 2002.

216. R. Leveque. *Finite volume methods for hyperbolic problems*. Cambridge University Press, Cambridge, UK, 2002.

217. T. Li. Exact boundary controllability of unsteady flows in a network of open canals. In *Differential equations & asymptotic theory in mathematical physics*, volume 2 of *Ser. Anal.*, pages 310–329. World Sci. Publ., Hackensack, NJ, 2004.

218. T. Li. Exact boundary controllability of unsteady flows in a network of open canals. *Math. Nachr.*, 278(3):278–289, 2005.

219. T. Li and L. Yu. Local exact boundary controllability of entropy solutions to linearly degenerate quasilinear hyperbolic systems of conservation laws. *ESAIM Control Optim. Calc. Var.*, 24(2):793–810, 2018.

220. T. T. Li. *Global classical solutions for quasilinear hyperbolic systems*, volume 32 of *RAM: Research in Applied Mathematics*. Masson, Paris; John Wiley & Sons, Ltd., Chichester, 1994.

221. Y. Li, E. Canepa, and C. Claudel. Optimal control of scalar conservation laws using linear/quadratic programming: Application to transportation networks. *IEEE Transactions on Control of Network Systems*, 1(1):28–39, 2014.

222. T. Liard and B. Piccoli. On entropic solutions to conservation laws coupled with moving bottlenecks. working paper or preprint, June 2019.

223. T. Liard and B. Piccoli. Well-posedness for scalar conservation laws with moving flux constraints. *SIAM J. Appl. Math.*, 79(2):641–667, 2019.

224. M. J. Lighthill and G. B. Whitham. On kinematic waves. II. A theory of traffic flow on long crowded roads. *Proc. Roy. Soc. London. Ser. A.*, 229:317–345, 1955.

225. P.-L. Lions. *Generalized solutions of Hamilton-Jacobi equations*, volume 69. London Pitman, 1982.

226. T. P. Liu. Invariants and asymptotic behavior of solutions of a conservation law. *Proc. Amer. Math. Soc.*, 71(2):227–231, 1978.

227. X.-Y. Lu, T. Z. Qiu, P. Varaiya, R. Horowitz, and S. E. Shladover. Combining variable speed limits with ramp metering for freeway traffic control. In *Proceedings of the 2010 American control conference*, pages 2266–2271. IEEE, 2010.

228. X.-Y. Lu, S. E. Shladover, I. Jawad, R. Jagannathan, and T. Phillips. Novel algorithm for variable speed limits and advisories for a freeway corridor with multiple bottlenecks. *Transportation Research Record: Journal of the Transportation Research Board*, 2489:86–96, 2015.

229. H. Ly, K. Mease, and E. Titi. Distributed and boundary control of the viscous Burgers equation. *Numerical Functional Analysis and Optimization*, 18(1):143–188, 1997.

230. A. Kesting. M. Treiber. *Traffic Flow Dynamics*. Springer-Verlag Berlin Heidelberg, 2013.

231. R. Manzo, B. Piccoli, and L. Rarità. Optimal distribution of traffic flows at junctions in emergency cases. *European Journal of Applied Mathematics*, 23(4):515–535, 2012.

232. A. Marigo and B. Piccoli. A fluid dynamic model for T-junctions. *SIAM J. Math. Anal.*, 39(6):2016–2032, 2008.

233. F. A. Mehmeti. *Nonlinear wave in networks*. Akademie Verlag, 1994.

234. K. Moskowitz and L. Newman. Notes on Freeway Capacity: Submitted for the Consideration of Committee on Highway Capacity, Highway Research Board. (4):72, 1961.

235. R. Nishi, A. Tomoeda, K. Shimura, and K. Nishinari. Theory of jam-absorption driving. *Transportation Research Part B: Methodological*, 50:116–129, 2013.

236. N. Risebro. P. Holden. *Front Tracking for Hyperbolic Conservation Laws*. Springer-Verlag Berlin Heidelberg, 2002.

237. M. Papageorgiou, H. Hadj-Salem, and J.-M. Blosseville. ALINEA: A local feedback control law for on-ramp metering. *Transportation Research Record*, 1320:58–64, 1991.

238. M. Papageorgiou, E. Kosmatopoulos, and I. Papamichail. Effects of variable speed limits on motorway traffic flow. *Transportation Research Record*, 2047(1):37–48, 2008.

239. C. Parzani and C. Buisson. Second-order model and capacity drop at merge. *Transportation Research Record: Journal of the Transportation Research Board*, 2315:25–34, 2012.

240. V. Perrollaz. Exact controllability of scalar conservation laws with an additional control in the context of entropy solutions. *SIAM Journal on Control and Optimization*, 50(4):2025–2045, 2012.

241. V. Perrollaz. Asymptotic stabilization of entropy solutions to scalar conservation laws through a stationary feedback law. *Ann. Inst. H. Poincaré Anal. Non Linéaire*, 30(5):879–915, 2013.

242. G. Piacentini, P. Goatin, and A. Ferrara. Traffic control via moving bottleneck of coordinated vehicles. *IFAC-PapersOnLine*, 51(9):13–18, 2018.

243. G. Piacentini, P. Goatin and A. Ferrara, A macroscopic model for platooning in highway traffic, *SIAM J. Appl. Math.*, 80(1):639–656, 2020.

244. C. Prieur, J. Winkin, and G. Bastin. Robust boundary control of systems of conservation laws. *Mathematics of Control, Signals, and Systems*, 20(2):173–197, 2008.

245. T. H. Qin. Global smooth solutions of dissipative boundary value problems for first order quasilinear hyperbolic systems. *Chinese Ann. Math. Ser. B*, 6(3):289–298, 1985. A Chinese summary appears in Chinese Ann. Math. Ser. A **6** (1985), no. 4, 514.

246. R. A. Ramadan and B. Seibold. Traffic flow control and fuel consumption reduction via moving bottlenecks. *arXiv preprint arXiv:1702.07995*, 2017.

247. J. Reilly, W. Krichene, M. L. Delle Monache, S. Samaranayake, P. Goatin, and A. M. Bayen. Adjoint-based optimization on a network of discretized scalar conservation law PDEs with applications to coordinated ramp metering. *Journal of optimization theory and applications*, 167(2):733–760, 2015.

248. J. Reilly, S. Martin, M. Payer, and A. Bayen. Creating complex congestion patterns via multi-objective optimal freeway traffic control with application to cyber-security. *Transportation Research Part B: Methodological*, 91:366–382, 2016. cited By 8.

249. P. I. Richards. Shock waves on the highway. *Operations Res.*, 4:42–51, 1956.

250. D. Robertson and R. Bretherton. Optimizing networks of traffic signals in real time—the scoot method. *IEEE Transactions on Vehicular Technology*, 40(1):11–15, 1991. cited By 357.

251. D. Serre. *Systèmes de lois de conservation. I.* Fondations. [Foundations]. Diderot Editeur, Paris, 1996. Hyperbolicité, entropies, ondes de choc. [Hyperbolicity, entropies, shock waves].

252. D. Serre. *Systèmes de lois de conservation. II.* Fondations. [Foundations]. Diderot Editeur, Paris, 1996. Structures géométriques, oscillation et problèmes mixtes. [Geometric structures, oscillation and mixed problems].

253. S. Shelby. Single-intersection evaluation of real-time adaptive traffic signal control algorithms. *Transportation Research Record*, 1867:183–192, 2004. cited By 26.

254. N. Shlayan and P. Kachroo. Feedback ramp metering using Godunov method based hybrid model. *Journal of Dynamic Systems, Measurement and Control, Transactions of the ASME*, 135(5), 2013. cited By 7.

255. M. Slemrod. Boundary feedback stabilization for a quasi-linear wave equation. In *Control Theory for Distributed Parameter Systems and Applications*, pages 221–237. Springer, 1983.

256. N. Smaoui. Boundary and distributed control of the viscous Burgers equation. *Journal of Computational and Applied Mathematics*, 182(1):91–104, 2005.

257. P. Spellucci. *Numerische Verfahren der Nichtlinearen Optimierung*. Birkhäuser-Verlag, Basel, 1993.

258. P. Spellucci. A new technique for inconsistent QP problems in the SQP method. *Math. Methods Oper. Res.*, 47(3):355–400, 1998.

259. P. Spellucci. An SQP method for general nonlinear programs using only equality constrained subproblems. *Mathematical Programming*, 82(3):413–448, 1998.

260. A. Srivastava and N. Geroliminis. Empirical observations of capacity drop in freeway merges with ramp control and integration in a first-order model. *Transportation Research Part C: Emerging Technologies*, 30:161–177, 2013.

261. R. E. Stern, Y. Chen, M. Churchill, F. Wu, M. L. D. Monache], B. Piccoli, B. Seibold, J. Sprinkle, and D. B. Work. Quantifying air quality benefits resulting from few autonomous vehicles stabilizing traffic. *Transportation Research Part D: Transport and Environment*, 67:351–365, 2019.

262. R. E. Stern, S. Cui, M. L. D. Monache, R. Bhadani, M. Bunting, M. Churchill, N. Hamilton, R. Haulcy, H. Pohlmann, F. Wu, B. Piccoli, B. Seibold, J. Sprinkle, and D. B. Work. Dissipation of stop-and-go waves via control of autonomous vehicles: Field experiments. *Transportation Research Part C: Emerging Technologies*, 89:205 –221, 2018.

263. Y. Sugiyama, M. Fukui, M. Kikuchi, K. Hasebe, A. Nakayama, K. Nishinari, S.-i. Tadaki, and S. Yukawa. Traffic jams without bottlenecks—experimental evidence for the physical mechanism of the formation of a jam. *New journal of physics*, 10(3):033001, 2008.

264. A. Talebpour and H. S. Mahmassani. Influence of connected and autonomous vehicles on traffic flow stability and throughput. *Transportation Research Part C: Emerging Technologies*, 71:143 –163, 2016.

265. B. Temple. Systems of conversation laws with invariant submanifolds. *Transactions of the American Mathematical Society*, 280(2):781–795, 1983.

266. E. Tomer, L. Safonov, N. Madar, and S. Havlin. Optimization of congested traffic by controlling stop-and-go waves. *Physical Review E - Statistical Physics, Plasmas, Fluids, and Related Interdisciplinary Topics*, 65(6), 2002. cited By 14.

267. M. Treiber and A. Kesting. *Traffic flow dynamics*. Springer, Heidelberg, 2013. Data, models and simulation, Translated by Treiber and Christian Thiemann.

268. J. Treiterer and J. Myers. The hysteresis phenomenon in traffic flow. In *Proceedings of the 6th International Symposium on Transportation and Traffic Theory, Sydney, Australia*, 1974.

269. N.-T. Trinh, V. Andrieu, and C.-Z. Xu. Design of integral controllers for nonlinear systems governed by scalar hyperbolic partial differential equations. *IEEE Trans. Automat. Control*, 62(9):4527–4536, 2017.

270. S. Ulbrich. A sensitivity and adjoint calculus for discontinuous solutions of hyperbolic conservation laws with source terms. *SIAM Journal on control and optimization*, 41(3):740–797, 2002.

271. S. Ulbrich. Adjoint-based derivative computations for the optimal control of discontinuous solutions of hyperbolic conservations laws. *Systems and control letters*, 48(3):313–328, 2003.

272. S. Villa, P. Goatin, and C. Chalons. Moving bottlenecks for the Aw-Rascle-Zhang traffic flow model. *Discrete Contin. Dyn. Syst. Ser. B*, 22(10):3921–3952, 2017.

273. M. Wang, W. Daamen, S. P. Hoogendoorn, and B. van Arem. Connected variable speed limits control and car-following control with vehicle-infrastructure communication to resolve stop-and-go waves. *Journal of Intelligent Transportation Systems*, 20(6):559–572, 2016.

274. M. Wang, W. Daamen, S. P. Hoogendoorn, and B. van Arem. Cooperative car-following control: Distributed algorithm and impact on moving jam features. *IEEE Transactions on Intelligent Transportation Systems*, 17(5):1459–1471, 2016.

275. F. Wu, R. E. Stern, S. Cui, M. L. D. Monache], R. Bhadani, M. Bunting, M. Churchill, N. Hamilton, R. Haulcy, B. Piccoli, B. Seibold, J. Sprinkle, and D. B. Work. Tracking vehicle trajectories and fuel rates in phantom traffic jams: Methodology and data. *Transportation Research Part C: Emerging Technologies*, 99:82 – 109, 2019.

276. Y. Xuan, V. Gayah, M. Cassidy, and C. Daganzo. Presignal used to increase bus-and car-carrying capacity at intersections. *Transportation Research Record*, 2315:191–196, 2012. cited By 16.

277. H. Zhang. A non-equilibrium traffic model devoid of gas-like behavior. *Transportation Research Part B: Methodological*, 36(3):275 – 290, 2002.

Index

© The Author(s) 2022
A. Bayen et al., *Control Problems for Conservation Laws with Traffic Applications*,
PNLDE Subseries in Control 99, https://doi.org/10.1007/978-3-030-93015-8

Printed in the United States
by Baker & Taylor Publisher Services